CRITICAL THINKING & LOGICAL REASONING WORKBOOK-0

GIFT OF LOGIC™ SERIES

Boost Your Thinking Skills

An Essential Resource for Everyone

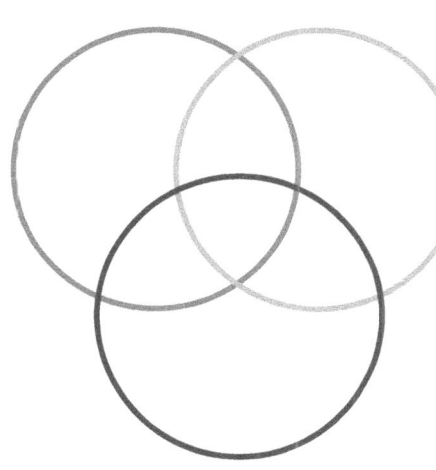

Verbal Reasoning

Analytical Reasoning

Pictorial Reasoning

THIRD EDITION

| FOR GRADES K-2 | STUDENTS, TEACHERS, AND PARENTS |

Ranga Raghuram

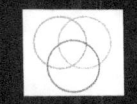 **GIFT OF LOGIC**™

 **Gift Of Logic, Inc**
http://www.giftoflogic.com
sales@giftoflogic.com

Critical Thinking and Logical Reasoning Workbook-0
ISBN-13: 978-1494822682
ISBN-10: 1494822687

Third Edition
1-2014

Copyright © 2009 Gift Of Logic, Inc. All rights reserved. No part of this publication may be reproduced, stored in a retrieval system, transmitted in any form or by any means, electronic, mechanical, photocopying, recording or otherwise, without the written permission of the publisher.

License: This book is licensed for use by one person only. Use of this book in a group setting (classroom, workshop, etc) without the written permission of the publisher is prohibited. Unauthorized duplication is strictly prohibited by law. Contact the publisher at sales@giftoflogic.com for classroom/school/group licensing.

GIFT OF LOGIC™
CRITICAL THINKING & LOGICAL REASONING CURRICULUM
12 WORKBOOKS TO BOOST YOUR THINKING SKILLS

	For Kindergarten, Grade 1, and Grade 2	
Workbook# 0 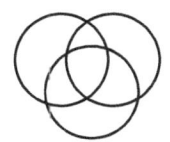	Verbal Reasoning	Finding the truth, Inferencing, Analogies, Synonyms and Antonyms, Agree/Disagree
	Analytic Reasoning	Memory drill, Decision making, Positioning, Sudoku
	Pictorial Reasoning	Connect the dots, Mazes, Picture Sequence, Spot the difference, etc
Workbook# 1 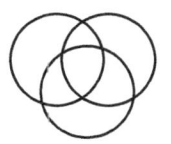	Verbal Reasoning	Finding the truth, Inferencing, Analogies, Synonyms and Antonyms, Agree/Disagree
	Analytic Reasoning	Sorting, Positioning, Picking, Assorted problems, Numeric and Alphabetic Sudoku
	Pictorial Reasoning	Picture Sequence, Spot the difference, Odd picture
Workbook# 2 	Verbal Reasoning	Finding the truth, Classification, Direct and Inverse relationship, Inferencing, Analogies, Agree/Disagree
	Analytic Reasoning	Sequencing, Scheduling, Strategy, Picking, etc
	Pictorial Reasoning	Picture Analogy, Odd picture, Pattern matching, etc
	For Grade 3, Grade 4, and Grade 5	
Workbook# 3 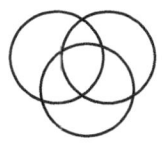	Verbal Reasoning	Not, And, Or, If .. then, Conditional inferencing, Unconditional inferencing, Symbolic Logic
	Analytic Reasoning	Lists, Sequencing, Grouping, Venn Diagrams, Graph logic, Number logic, Letter logic, Sudoku
	Pictorial Reasoning	Picture sequence, Picture analogy, Odd picture, Picture difference, Pattern matching
Workbook# 4 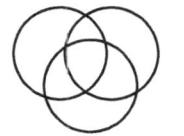	Verbal Reasoning	Contradiction, Converse, Inverse, Contrapositive, Conditional inferencing, Symbolic Logic
	Analytic Reasoning	Scheduling, Looping, FIFO, LIFO, Correlation, Venn Diagram, Graph logic, Number logic, Sudoku, etc
	Pictorial Reasoning	Picture sequence, Picture analogy, Odd picture, Picture difference, Pattern matching
Workbook# 5 	Verbal Reasoning	Biconditional, Categorical inferencing, Cause and Effect, Symbolic Logic, Agree/Disagree, Word and Sentence analogy
	Analytic Reasoning	Correlation, Grouping, Venn Diagrams, Graph logic, Number logic, Letter logic, Sudoku, etc
	Pictorial Reasoning	Picture sequence, Picture analogy, Odd picture, Picture difference, Pattern matching

********* Essential resource for everyone *********

*http://www.giftoflogic.com   *sales@giftoflogic.com

GIFT OF LOGIC™
CRITICAL THINKING & LOGICAL REASONING CURRICULUM
12 WORKBOOKS TO BOOST YOUR THINKING SKILLS

For Grades 6-12, College/University Students, Adults

Primer

Prereq

Verbal Reasoning	Logical Operators, Conditional, Categorical and Causal reasoning, Validity, Fallacies, Symbolic Logic
Analytic Reasoning	Positioning, Grouping, Sudoku
Pictorial Reasoning	Pattern perception, Figure formation, Paper folding and cutting, Figure matrix, Rule detection

Workbook# 6

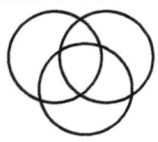

Verbal Reasoning	Arguments-Main point, Must be true, Cannot be true
Analytic Reasoning	Positioning, Grouping, Sudoku
Pictorial Reasoning	Pattern perception, Figure formation, Paper folding and cutting, Figure matrix, Rule detection

Workbook# 7

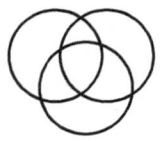

Verbal Reasoning	Arguments-Strengthening, Weakening
Analytic Reasoning	Positioning, Grouping, Sudoku
Pictorial Reasoning	Pattern perception, Figure formation, Paper folding and cutting, Figure matrix, Rule detection

Workbook# 8

Verbal Reasoning	Arguments - Controversy, Paradox
Analytic Reasoning	Positioning, Grouping, Sudoku
Pictorial Reasoning	Pattern perception, Figure formation, Paper folding and cutting, Figure matrix, Rule detection

Workbook# 9

Verbal Reasoning	Arguments- Assumptions, Reasoning strategy
Analytic Reasoning	Positioning, Grouping, Sudoku
Pictorial Reasoning	Pattern perception, Figure formation, Paper folding and cutting, Figure matrix, Rule detection

Workbook# 10

Verbal Reasoning	Arguments-Flawed reasoning, Analogous reasoning
Analytic Reasoning	Positioning, Grouping, Sudoku
Pictorial Reasoning	Pattern perception, Figure formation, Paper folding and cutting, Figure matrix, Rule detection

********* Essential resource for everyone *********
Get the GIFT OF LOGIC™ today !
*http://www.giftoflogic.com *sales@giftoflogic.com

Dear Reader:

Your decision to purchase this book is commendable. You now have in your hands, a comprehensive, easy-to-read book in Critical thinking and Logical reasoning that will introduce you to three different areas of thinking and reasoning - Verbal, Analytical and Pictorial. Solving problems in Verbal Reasoning is important to develop a critical mind. Solving problems in Analytic Reasoning is important to develop a flexible and resourceful mind. Solving problems in Pictorial Reasoning is important to develop a visually alert mind.

This book is presented in a workbook format to help you progress quickly. Parents and teachers are urged to complete the exercises ahead of the student and assist them whenever necessary with the help of detailed answers provided at the end of the book. This book can be used as a supplementary resource in the regular class room or it can be used during winter and summer vacations. College/University students, working professionals and retired individuals will also find the Gift Of Logic(tm) Series very useful in enhancing their problem solving abilities, confidence and general intellect.

Critical thinking and Logical reasoning must be practiced consistently to develop strong cognitive skills. After completing the exercises in this book, continue to read the other books in this series to get familiar with different types of Logical reasoning problems.

This workbook is one in a series of twelve workbooks. Please refer to the brochure before this page for a brief description of each workbook. Visit the website http://www.giftoflogic.com for more information.

                                                     Happy thinking and reasoning!

TABLE OF CONTENTS

Verbal Reasoning

Finding the truth..9

Word Analogy..11

Synonyms/Antonyms...13

Agree or Disagree..15

Inferencing (Deductive Reasoning) 17
    Who.. 18
    Where 19
    When... 20
    Classification 21

Analytic Reasoning

Memory drill..23

Decision making..25

Positioning.. 33

Sudoku
    Numeric Sudoku... 38
    Alphabetic Sudoku...48

TABLE OF CONTENTS

Pictorial Reasoning

Connect the dots..59

Mazes.. 64

Picture Sequence...68

Odd Picture.. 72

Spot the difference..75

Answers

Verbal...79

Analytic..87

Pictorial..111

Certificate of Completion

Name _____   Date _____

VERBAL REASONING

Name —————————— Date ——————————

FINDING THE TRUTH

Finding the truth of statements is important for reasoning correctly. Find the truth of the following statements and circle the correct answer.

1	Every year has twelve months. A) True B) False
2	Every month has thirty days. A) True B) False
3	Only airplanes can travel in the air. A) True B) False
4	Dinosaurs lived on earth millions of years ago. A) True B) False
5	Only dogs can be kept as pets. A) True B) False
6	We can see things with our eyes closed. A) True B) False

Verbal Reasoning

© Gift Of Logic, Inc * Copying prohibited

Name _____   Date _____

FINDING THE TRUTH

Finding the truth of statements is important for reasoning correctly.
Find the truth of the following statements and circle the correct answer.

7	Eiffel Tower is in the city of London in United Kingdom. A) True B) False
8	Pluto is not a planet in the Solar system. A) True B) False
9	New York is near the Pacific Ocean. A) True B) False
10	Some car drivers are men. A) True B) False
11	All car drivers are men. A) True B) False
12	Fire can be extinguished (put out) with water. A) True B) False

Verbal Reasoning   Answers-79
© Gift Of Logic, Inc * Copying prohibited

Name _____   Date _____

WORD ANALOGY

The first two words separated by a colon (:) have a specific relationship. The third word has the same relationship with one of the words in the answer choices given. Circle the answer choice that will complete the analogy.

Example:

   tire: wheel ==> nail :
          A) hair   B) finger

The relationship between a tire and a wheel can be described by the phrase "attached to". A tire is attached to a wheel. Similarly, a nail is attached to a finger. So, B is the correct answer.

1	bird : wings => fish : A) gills B) fins
2	electricity : wire => water : A) tap B) pipe
3	chair : sit => treadmill : A) exercise B) sleep
4	oven : hot => fridge : A) milk B) cool

Name _____   Date _____

WORD ANALOGY

The first two words separated by a colon (:) have a specific relationship. The third word has the same relationship with one of the words in the answer choices given. Circle the answer choice that will complete the analogy.

5	class : study => park : A) play B) worship
6	room : door => box : A) lid B) key
7	toothache : dentist => fever : A) doctor B) lawyer
8	success : win => defeat : A) lose B) happy
9	increase : raise => decrease : A) reduce B) stop
10	eye : see => ear : A) taste B) hear

Verbal Reasoning   Answers-80
© Gift Of Logic, Inc * Copying prohibited

Name _____   Date _____

SYNONYMS/ANTONYMS

Synonyms are words with the same meaning. Antonyms are words with opposite meaning. The following statements are made using synonyms or antonyms. Read each statement and find out if it is true or false.

1	Being noisy is the same as being quiet. A) True B) False
2	Being rude is the same as being arrogant. A) True B) False
3	To be slim is the same as being thin. A) True B) False
4	To be fat is the same as being heavy. A) True B) False
5	To be friends with someone is the same as to quarrel with someone. A) True B) False
6	To clean the mess is not the same as keeping things tidy. A) True B) False
7	To elevate something is the same as bringing it down. A) True B) False

Verbal Reasoning
© Gift Of Logic, Inc * Copying prohibited

Name ——————————————   Date——————————————

SYNONYMS/ANTONYMS

Synonyms are words with the same meaning. Antonyms are words with opposite meaning. The following statements are made using synonyms or antonyms. Read each statement and find out if it is true or false.

8	To scatter the flowers is the same as gathering the flowers. A) True B) False
9	To help someone is the same as hindering them. A) True B) False
10	Permit is the opposite of stop. A) True B) False
11	To be happy is not the same as being sad. A) True B) False
12	To polish a shoe is not the same as making it shine. A) True B) False
13	To be courteous is the same as being rude. A) True B) False
14	To hate something is not the same as liking it. A) True B) False

Verbal Reasoning   Answers-81
© Gift Of Logic, Inc * Copying prohibited

Name _____ Date _____

AGREE-DISAGREE

Read the statements of two people and find out whether they both agree or disagree with each other.

1

Victor: Apples taste better than Grapes.

Victoria: Grapes taste better than apples.

Victor and Victoria
 A) agree with each other.
 B) disagree with each other.

2

Steve: Ants are smarter than elephants.

Emily: Elephants are smarter than ants.

Steve and Emily
 A) agree with each other.
 B) disagree with each other.

Name ――――――――――――――   Date――――――――――――――

AGREE-DISAGREE

Read the statements of two people and find out whether they both agree or disagree with each other.

3

Daniel: When Josh got hurt, the ambulance was called. The ambulance came quickly and the nurse applied a bandage.

Donna: The ambulance carrying the nurse could not find us quickly. After she arrived, the nurse applied the bandage.

Daniel and Donna
    A) agree that the ambulance arrived quickly.
    B) disagree that the ambulance arrived quickly.

4

Mom: Tony likes reading more than sports.
Dad: If Tony has to choose between reading and sports, he will choose reading.

Mom and Dad
    A) agree with each other.
    B) disagree with each other.

Name —————————————— Date——————————

INFERENCING

Inferencing means to find facts that are not stated. In the following questions, some facts are given. Assuming that these facts are true, you must find if any other fact can be inferred. Use your common sense to answer these questions.

1 before/after

Tom was late for the 6 O'Clock party, but Ryan was not.

We can conclude from the above statement that
    A) Tom went to the party before Ryan.
    B) Ryan went to the party before Tom.

2 location

The diamond ring was in the jewel case. The jewel case was found in the trash can.

We can conclude from the above statements that
    A) Trash was found in the jewel case.
    B) The diamond ring was found in the trash can.

3 or

Amy's mom takes her to the park on Mondays and Wednesdays, and her dad takes her to the park on other days. Amy is in the park now, and today is a Tuesday.
We can infer from the above statements that
    A) Amy's mom took her to the park today.
    B) Amy's dad took her to the park today.

Name _____   Date _____

INFERENCING - Who

4

Brian chased a thief for a few miles, arrested him, and took him to jail.

We can infer from the above statement that Brian is most likely
    A) a Doctor
    B) an Engineer
    C) a Police Officer

5

As Stacy stood in front of the camera and read the drama script, tears rolled from her eyes.

Stacy is most likely
    A) an actress
    B) a dentist

6

A bridge was under construction in the city. James gave instructions to the workers to pour the concrete correctly.

James is most likely
    A) a Doctor
    B) a Lawyer
    C) an Engineer

Verbal Reasoning
© Gift Of Logic, Inc * Copying prohibited

Name _____   Date _____

INFERENCING - Where

7

Troy dived several hundred feet. He was looking for his golden watch.

Troy is most likely in a
- A) a shallow pond
- B) a deep lake
- C) a bathtub

8

John was riding in his speedboat at night when it suddenly drowned. After swimming to the shore, he felt tired and slept under a tree. When he woke up in the morning, there was nobody around him.

John was most likely in
- A) a remote place
- B) the center of a city

9

Neil jumped and ended up with broken bones and head injuries.

Neil is likely to have jumped from
- A) a tall building
- B) a small chair

Verbal Reasoning   Answers-84
© Gift Of Logic, Inc * Copying prohibited

Name ———————————————— Date————————————

INFERENCING - When

10 **before**

Martha watched TV in the evening until 6 PM. Then, she went to her friend's house. When she returned home after two hours, to her surprise, the TV was stolen.

Based on the above statements, we can infer that her TV was stolen
　　A) before 6 PM or after 8 PM.
　　B) between 6 PM and 8 PM.

11 **after/before**

The Police rushed to the train station to catch a robber in a train. But, by the time they arrived at the station, the train had already left.

Based on the above statements, we can infer that the Police arrived
　　A) after the robber had left the train station.
　　B) before the robber had left the train station.

12 **at the same time**

Jack and Jill were at the dental office at 8 AM. It would take a dentist thirty minutes to clean one person's teeth. Jack and Jill got their teeth cleaned, and left the Dentist at 8:30 AM.
Based on the above information, we can infer that
　A) Jack's teeth was cleaned first and then Jill's.
　B) Jack and Jill got their teeth cleaned at the same time.

Verbal Reasoning　　　　　　　　　Answers-85
© Gift Of Logic, Inc * Copying prohibited

Name _____ Date _____

INFERENCING - Classification

13 some

Some wooden homes are big, some are small. Some brick homes are big, some are small.

Based on the above statements, we can infer that

1) If a home is big, it must be a wooden home.
   A) True   B) False

2) If a home is made of bricks, it must be small.
   A) True   B) False

---

14 some

Some people are doctors, and some people are nurses. Doctors are either male or female. So are nurses.

Which one of the following can be inferred from the above facts?
  1) All doctors are female.
     A) True   B) False

  2) All nurses are male.
     A) True   B) False

  3) All females are doctors.
     A) True   B) False

Verbal Reasoning
© Gift Of Logic, Inc * Copying prohibited

ANALYTICAL REASONING

Name ———————————   Date ———————————

MEMORY DRILL

Remembering numbers and alphabets in reverse order will improve your memory retention skills.

1

Think of any three numbers and retain them in your memory. The numbers should not be consecutive. Write them in reverse order in the table below.

Think of a new set of three non-consecutive numbers and write them in reverse order below.

Repeat this exercise one more time with another set of three numbers.

2
Think of any three numbers and retain these numbers in your memory.
Add two to the third number and write it in the first block below.
Add one to the first number and write it in the second block.
Subtract one from the second number and write it in the third block.

3
Think of any three numbers and retain these numbers in your memory.

Add one to the first number and write it in the first block.
Write the second number in the second block.
Subtract one from the third number and write it in the third block.

Analytical Reasoning
© Gift Of Logic, Inc * Copying prohibited

MEMORY DRILL

4

a	f
4	9

Read the two rows of the table a few times and retain it in your memory. Remember what is in the first row and the second row. Now, hide the table with your hand and answer the following questions.

A) What is the second letter in the first row?
B) What is the first number in the second row?
C) What is the first letter in the first row?

5

Read the rows a few times and retain them in memory.

c	e
4	k
k	c

Now, hide the table with your hand and answer the following questions.

A) How many c's are there?
B) How many e's are there?
C) How many k's are there?

Analytical Reasoning
© Gift Of Logic, Inc * Copying prohibited

Name —————————— Date ——————————

DECISION MAKING

In decision making problems that follow, read the information given in each problem and answer the question that follows. These type of problems sharpen your critical thinking ability and common sense.

1

which way?

It would take ten minutes to go to the Red river if you take the Pine Street. It would take longer to go to the river if you take the Elm Street.

Which street should one take to go to the Red river quickly?
   A) Pine Street
   B) Elm Street

2

when to sleep?

The sooner you go to bed tonight, the better you will do in your test.

When is the best time to go to bed tonight?

   A) 10 PM
   B) 8 PM

Name _____   Date _____

DECISION MAKING

3

add or remove?

The red milk carton has less milk than the blue milk carton. Both the milk cartons are of the same size. To make the milk levels the same in both cartons, what should we do?

    A) add more milk to the blue milk carton.
    B) add more milk to the red milk carton.

4

increase or decrease?

The baby elephant at the zoo weighs 200 pounds. The zoo keeper thought that it was a bit overweight. He wanted the elephant to weigh only 150 pounds.

What should the baby elephant do?

    A) put on more weight.
    B) lose some pounds.

Name ―――――――――――― Date ――――――――――――

DECISION MAKING

5

more or less?

During the rainy season, people buy extra food in order to avoid frequent trips to the stores. So, during the rainy season, store keepers must do which one of the following?

    A) keep less food in the shelves.
    B) keep more food in the shelves.

6

add or remove?

Kevin had saved 10 dollars in his piggy bank. He spent 5 dollars on movies. To bring his savings back to 10 dollars what should be done?

    A) More money must be added to the piggy bank.
    B) Money must be removed from the piggy bank.

Name ——————————————— Date ———————————————

DECISION MAKING

7

keep or return?

Anita paid three dollars to a shopkeeper for bread that costs two dollars.

What should the shopkeeper do?
   A) Keep the three dollars to himself.
   B) Return one dollar to Anita.

8

correct exchange?

There are ten students in Mr. Gary's class. There are eight students in Mr. Roger's class. To have the same number of students in both classes, what must be done?

A) Move one student from Mr. Roger's class to Mr. Gary's class.
B) Move one student from Mr. Gary's class to Mr. Roger's class.

Name ——————————   Date ——————————

DECISION MAKING

9

when to begin?

Silvia has to attend a party at 10 AM. But, before going to the party, she must complete her homework, which will take 30 minutes to do.

When should Silvia begin her homework?
    A) before 9:30 AM.
    B) after 9:30 AM.

10

when to start?   **July 29 - July 30 - July 31 - August 1**

The Science project must be submitted on the first of August. It will take Laura three days to complete the project.

How soon should Laura start working on her Science project?

    A) on or before July 29.
    B) any day after July 29.

DECISION MAKING

11

what to do?

To be fully occupied for one hour, which one of the following is the correct thing to do?

A) Work on five tasks, each of 10 minutes duration.
B) Work on four 10 minute tasks and one 20 minute task.

12

what to do?

The short hand of a clock was at 3, and the long hand of the clock was at 1, when the time was 3 PM. To correct the clock, what should be done?

A) The long hand must be moved back.
B) The short hand must be moved back.

Name ———————————— Date ————————————

DECISION MAKING

13

when to change?

Gina changes her pillow cover every other day.

1) If she changed her pillow cover on Wednesday, when will she change it again?
   A) Monday   B) Friday

2) If she changed her pillow cover on Wednesday this week, will she change it on Wednesday next week?
   A) Yes   B) No

14

what to do?

A school must conduct three fire drills from 9 AM to 11 AM, with a gap of one hour between drills.

When should the first fire drill begin?

A) after 9 AM.
B) at 9 AM.

Name _____   Date _____

DECISION MAKING

15

when to turn on?

A special red bulb will not glow immediately, but only fifteen minutes after it is turned on. If the bulb must glow at 10 AM, when should it be turned on?

    A) 9:45 AM   B) 10:00 AM   C) 10:15 AM

16

what to send?

A horse must be sent to the stage when the circus begins, and every five minutes thereafter. An elephant must also be sent to the stage when the circus begins, but every ten minutes thereafter. Fifteen minutes after the circus begins, what should be sent to the stage?

    A)   An elephant only.
    B)   An elephant and a horse.
    C)   A horse only.

Analytical Reasoning
© Gift Of Logic, Inc * Copying prohibited

Name _____   Date _____

POSITIONING   MATCHING

Solving positioning problems makes your brain flexible and resourceful. You will learn to find solutions even when there are restrictions.

1

| Red box | Green box | Blue box |

Three boxes shown above can hold balls of the colors that are shown in the box. Six balls of the colors shown below must be placed in the boxes.

Ball 1 - Green
Ball 2 - Red
Ball 3 - Blue
Ball 4 - Green
Ball 5 - Red
Ball 6 - Green

Answer the questions below.

1) How many balls can be placed in the red box?
   A) 2   B) 3

2) How many balls can be placed in the green box?
   A) 3   B) 6

3) How many balls can be placed in the blue box?
   A) 1   B) 2

POSITIONING — SAME SPOT

2

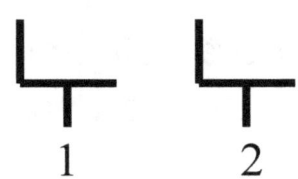

Jack and Jill must sit on the chairs shown above. Jack must sit in chair # 1.

Which one of the following seatings is correct?

A)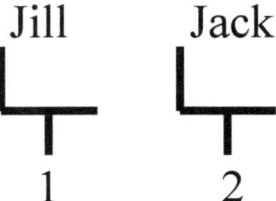

B)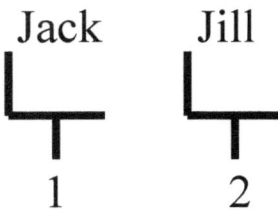

POSITIONING OR

3

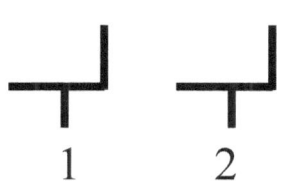

Mary and Nancy must be seated in the two chairs shown above. Mary can sit in either chair #1 or chair #2.

In the following seating arrangements, write the name of the person sitting in chair #1 and chair #2 next to the question mark.

A)

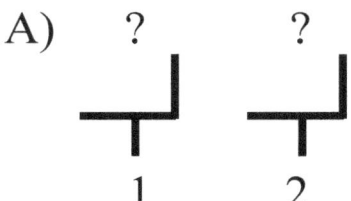

B)

POSITIONING   **NEXT TO**

4

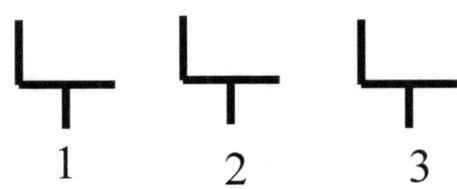

Rick, Shelly, and Tina must be seated in the three chairs shown above. Shelly must sit next to Rick. Tina must sit in chair # 3.

Write the names of the persons who will sit in the three chairs next to the question mark shown. In how many ways can they be seated?

A)   ?   ?   Tina
       1   2   3

B)   ?   ?   Tina
       1   2   3

C)   ?   ?   Tina
       1   2   3

Analytical Reasoning

Name ——————————————— Date ———————————

| POSITIONING | FIRST, LAST, ONLY |

5

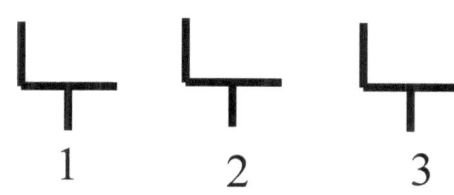

Emma, Frank, and Gary must be seated in the three chairs shown.
Frank can sit in the first chair or the last chair only.
If Frank sits on the first chair, where can Emma and Gary sit? Write their names in their possible positions below.

A)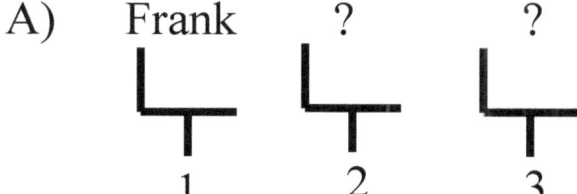

B)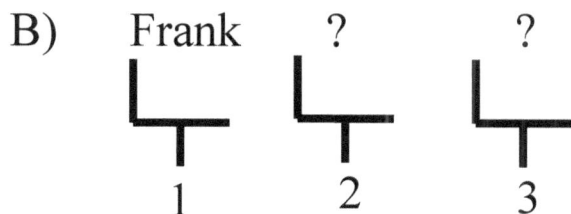

If Frank sits on the last chair, where can Emma and Gary sit? Write their names in their possible positions below.

A)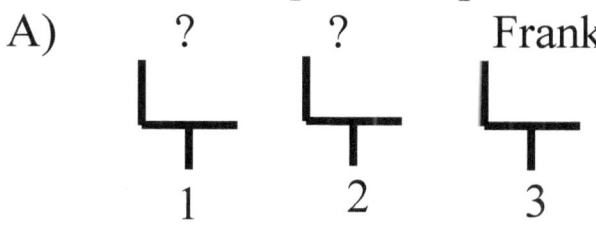

B)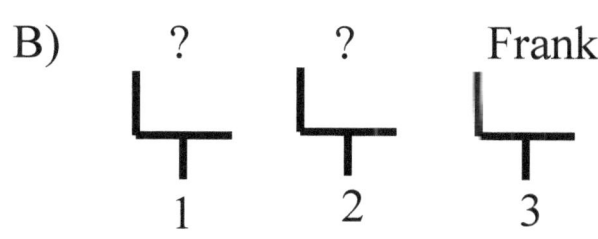

NUMERIC SUDOKU

Solve the Sudokus shown below. A solved Sudoku has numbers 1,2,3, and 4 appearing in each row, each column and the four bolded squares only once. You develop valuable positioning skills while solving these Sudokus.

1

	1	4	3
3		1	2
1	3		4
4	2	3	

2

1			2
2	3	1	4
4			3
3	2	4	1

Name _____   Date _____

NUMERIC SUDOKU

Solve the Sudokus shown below. A solved Sudoku has numbers 1,2,3, and 4 appearing in each row, each column and the four bolded squares only once. You develop valuable positioning skills while solving these Sudokus.

3

2			3
1	3	2	4
3			1
4	1	3	2

4

	3	1	
2	1	3	4
	4	2	
3	2	4	1

NUMERIC SUDOKU

Solve the Sudokus shown below. A solved Sudoku has numbers 1,2,3, and 4 appearing in each row, each column and the four bolded squares only once. You develop valuable positioning skills while solving these Sudokus.

5

1	3	4	2
	2	1	
	4	3	
3	1	2	4

6

	3		1
4		3	
1	4	2	3
3	2	1	4

NUMERIC SUDOKU

Solve the Sudokus shown below. A solved Sudoku has numbers 1,2,3, and 4 appearing in each row, each column and the four bolded squares only once. You develop valuable positioning skills while solving these Sudokus.

7

	2	3	1
3	1		2
1		2	3
2		1	

8

3	1	4	2
4			3
2	4		1
1	3	2	

NUMERIC SUDOKU

Solve the Sudokus shown below. A solved Sudoku has numbers 1,2,3, and 4 appearing in each row, each column and the four bolded squares only once. You develop valuable positioning skills while solving these Sudokus.

9

	4	1	3
1		2	4
3		4	1
4	1		2

10

4	3	1	2
2		3	4
1	4		3
3	2	4	

NUMERIC SUDOKU

Solve the Sudokus shown below. A solved Sudoku has numbers 1,2,3, and 4 appearing in each row, each column and the four bolded squares only once. You develop valuable positioning skills while solving these Sudokus.

11

4	1	3	
3	2		4
2		4	1
	4	2	3

12

1	2	4	
3	4		2
4		2	1
2		3	4

Name _____   Date _____

NUMERIC SUDOKU

Solve the Sudokus shown below. A solved Sudoku has numbers 1,2,3, and 4 appearing in each row, each column and the four bolded squares only once. You develop valuable positioning skills while solving these Sudokus.

13

4	1	2	3
2	3		1
3		1	2
	2	3	4

14

2		1	3
3		2	4
1	3		2
4	2		1

Analytical Reasoning

NUMERIC SUDOKU

Solve the Sudokus shown below. A solved Sudoku has numbers 1,2,3, and 4 appearing in each row, each column and the four bolded squares only once. You develop valuable positioning skills while solving these Sudokus.

15

4	1		3
3	2		4
2	3	4	
1	4	3	

16

1	3		4
	4	3	1
4	2		3
3	1	4	

Name _____   Date _____

NUMERIC SUDOKU

Solve the Sudokus shown below. A solved Sudoku has numbers 1,2,3, and 4 appearing in each row, each column and the four bolded squares only once. You develop valuable positioning skills while solving these Sudokus.

17

1	2	3	4
	4	1	2
4		2	1
2	1		3

18

4	1	3	
2			1
1	4		3
	2	1	4

Analytical Reasoning
© Gift Of Logic, Inc * Copying prohibited

NUMERIC SUDOKU

Solve the Sudokus shown below. A solved Sudoku has numbers 1,2,3, and 4 appearing in each row, each column and the four bolded squares only once. You develop valuable positioning skills while solving these Sudokus.

19

2	1	4	3
4		2	1
3	2		4
1		3	2

20

	3	2	1
1	2		3
2	1		4
3	4	1	

ALPHABETIC SUDOKU

Solve the Sudokus shown below. A solved Sudoku has alphabets A, B, C, and D appearing in each row, each column and the four bolded squares only once. You develop valuable positioning skills while solving these Sudokus.

1

	A	D	C
C		A	B
A	C		D
D	B	C	

2

A	D	C	
B	C		D
D		B	C
	B	D	A

ALPHABETIC SUDOKU

Solve the Sudokus shown below. A solved Sudoku has alphabets A, B, C, and D appearing in each row, each column and the four bolded squares only once. You develop valuable positioning skills while solving these Sudokus.

3

	D		C
A	C	B	D
	B		A
D	A	C	B

4

	C	A	B
B	A		D
	D		C
C	B	D	A

ALPHABETIC SUDOKU

Solve the Sudokus shown below. A solved Sudoku has alphabets A, B, C, and D appearing in each row, each column and the four bolded squares only once. You develop valuable positioning skills while solving these Sudokus.

5

	C	D	
D	B	A	C
B	D	C	A
	A	B	

6

B		D	
D	A	C	B
A		B	
C	B	A	D

ALPHABETIC SUDOKU

Solve the Sudokus shown below. A solved Sudoku has alphabets A,B,C, and D appearing in each row, each column and the four bolded squares only once. You develop valuable positioning skills while solving these Sudokus.

7

	B		A
C	A	D	B
A	D	B	C
	C		D

8

A	D	B	C
B		A	D
D	B		A
C		D	B

ALPHABETIC SUDOKU

Solve the Sudokus shown below. A solved Sudoku has alphabets A, B, C, and D appearing in each row, each column and the four bolded squares only once. You develop valuable positioning skills while solving these Sudokus.

9

		A	C
A	C		
C	B		A
D	A	C	B

10

D	C	A	B
	A	C	D
A			C
C	B		

Analytical Reasoning

ALPHABETIC SUDOKU

Solve the Sudokus shown below. A solved Sudoku has alphabets A, B, C, and D appearing in each row, each column and the four bolded squares only once. You develop valuable positioning skills while solving these Sudokus.

11

	A	C	B
C		A	D
B			A
A	D	B	

12

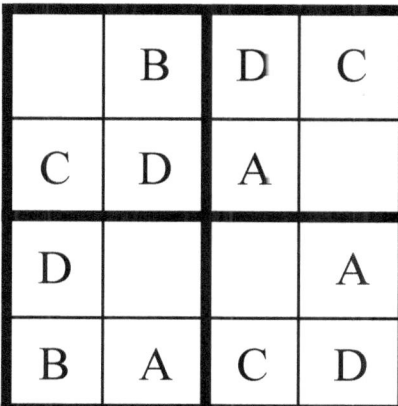

	B	D	C
C	D	A	
D			A
B	A	C	D

ALPHABETIC SUDOKU

Solve the Sudokus shown below. A solved Sudoku has alphabets A, B, C, and D appearing in each row, each column and the four bolded squares only once. You develop valuable positioning skills while solving these Sudokus.

13

C	D	B	A
B		D	
D	C	A	B
A		C	

14

	D	A	C
C	A		D
	C	D	B
D	B		A

Analytical Reasoning

ALPHABETIC SUDOKU

Solve the Sudokus shown below. A solved Sudoku has alphabets A, B, C, and D appearing in each row, each column and the four bolded squares only once. You develop valuable positioning skills while solving these Sudokus.

15

	B	A	
C	A	B	D
B		D	A
A	D		B

16

A	C	B	
	D		A
D	B	A	C
	A	D	

ALPHABETIC SUDOKU

Solve the Sudokus shown below. A solved Sudoku has alphabets A,B,C, and D appearing in each row, each column and the four bolded squares only once. You develop valuable positioning skills while solving these Sudokus.

17

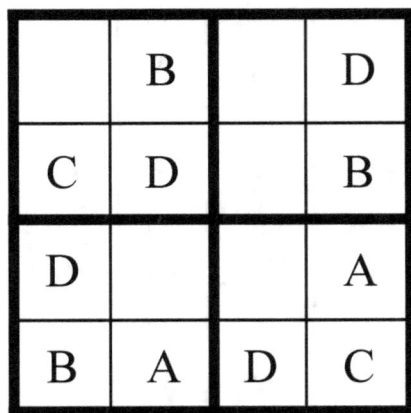

18

ALPHABETIC SUDOKU

Solve the Sudokus shown below. A solved Sudoku has alphabets A, B, C, and D appearing in each row, each column and the four bolded squares only once. You develop valuable positioning skills while solving these Sudokus.

19

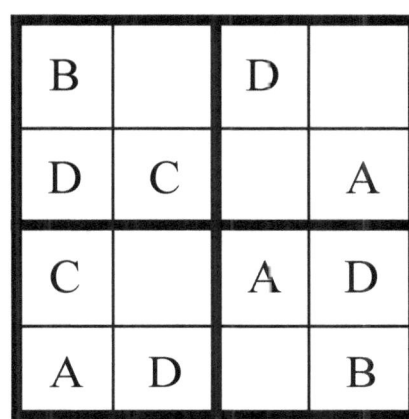

20

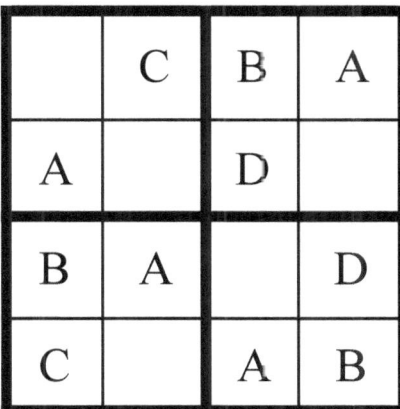

Name ———————————————   Date ———————————————

PICTORIAL REASONING

Name _____   Date _____

CONNECT THE DOTS

Connect the dots shown below, starting from the lowest number to the next number, and ending at the highest number. When you reach the highest number, connect it to the lowest number to complete the shape. What is the difference between the shape on the left and the shape on the right?

1
1 •   • 2 

4 •   • 3

2
           • 2 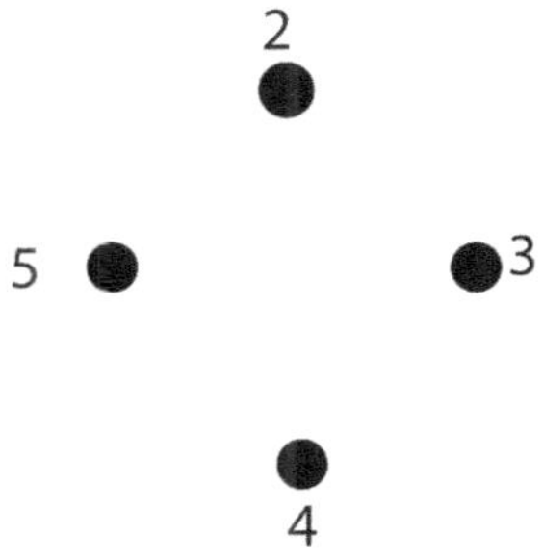
    5 •   • 3
           • 4

3
        • 2 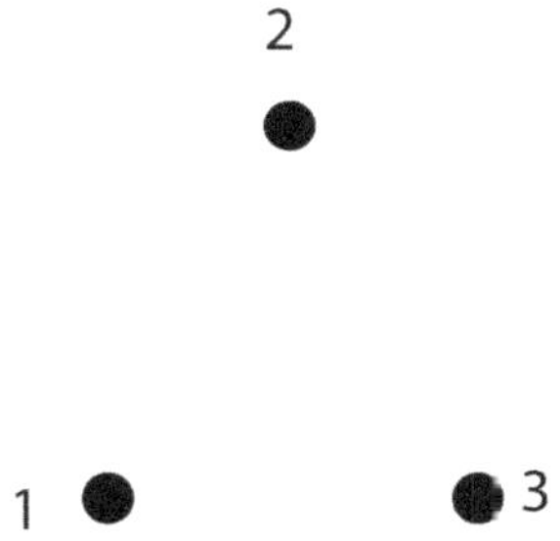

1 •   • 3

4
   • 2   • 3

      1 •

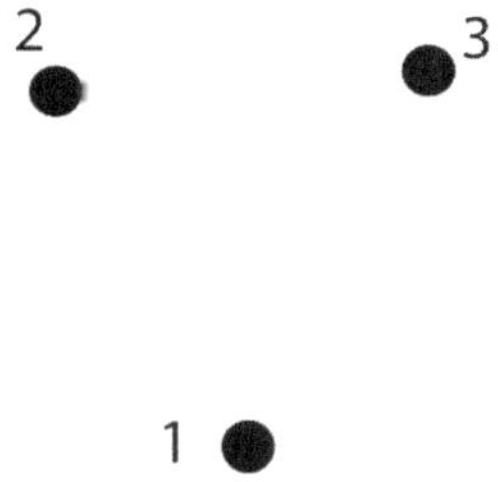

5
1 •   • 4
2 •   • 3

6 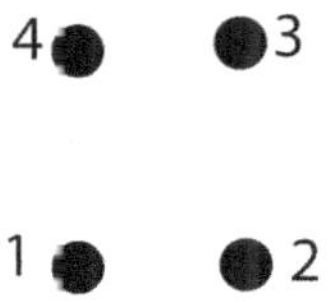
4 •   • 3
1 •   • 2

Pictorial Reasoning   Answers-111   59
© Gift Of Logic, Inc * Copying prohibited

Name —————————————— Date ——————————————

CONNECT THE DOTS

Connect the dots shown below, starting from the lowest number to the next number, and ending at the highest number. When you reach the highest number, connect it to the lowest number to complete the shape. What is the difference between the shape on the left and the shape on the right?

7
1 •   • 2

4 •   • 3

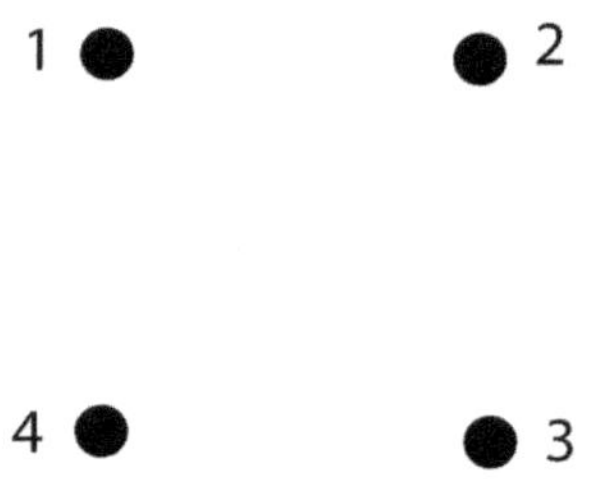

8
2 •   • 3

1 •   • 4

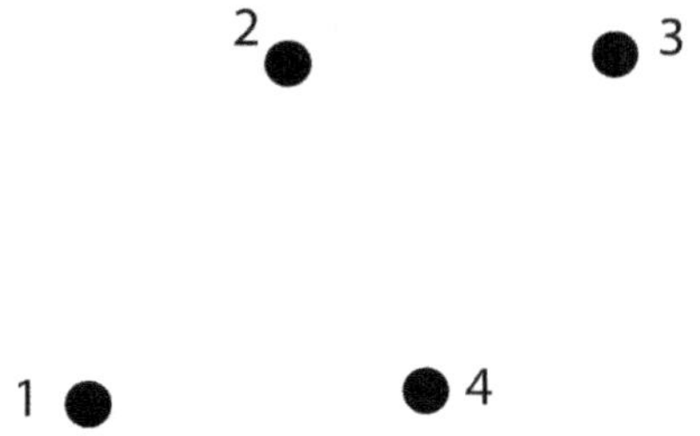

9
2 •   • 4

1 •   • 5
   • 3

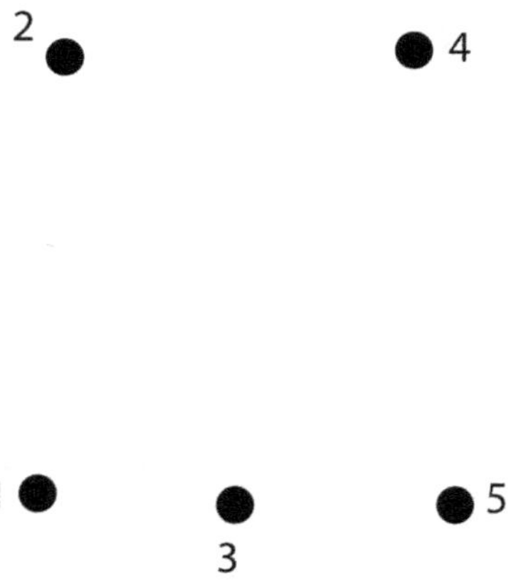

10
1 •   3 •   • 5

2 •   • 4

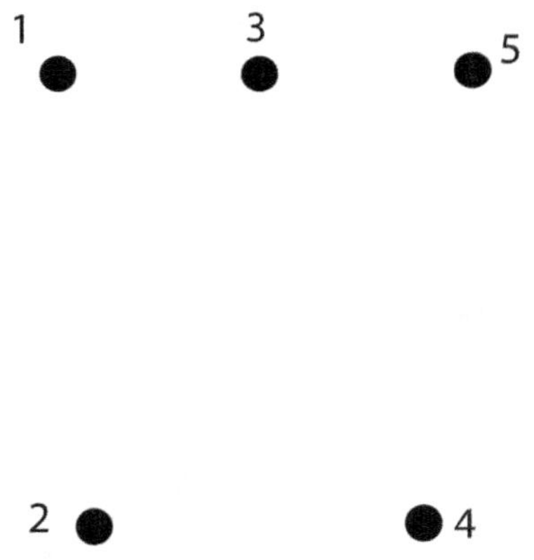

11
1 •

2 •   • 3

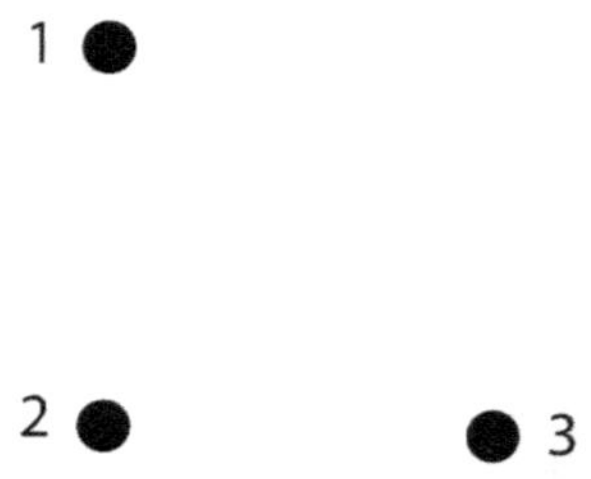

12 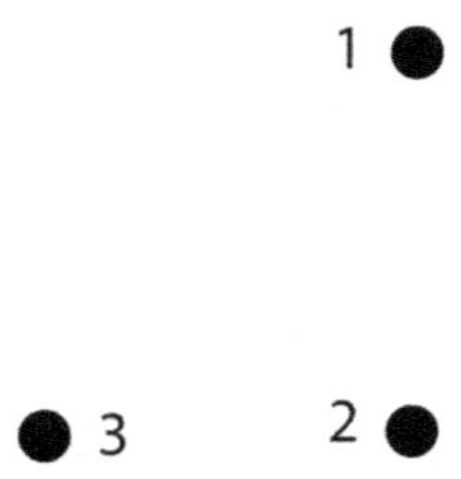
1 •

• 3   2 •

Pictorial Reasoning   Answers-111   60
© Gift Of Logic, Inc * Copying prohibited

Name _____   Date_____

CONNECT THE DOTS -JUMP

Connect the dots shown below, starting from the lowest number to the next number until you reach the highest number. Stop when you reach the highest number. The letter J is an instruction to jump to the number shown next to it without drawing a line to it. For example, 2-J-3 means, jump from 2 to 3 without drawing a line from 2 to 3. After jumping, continue to the next higher number that has not been connected.

13

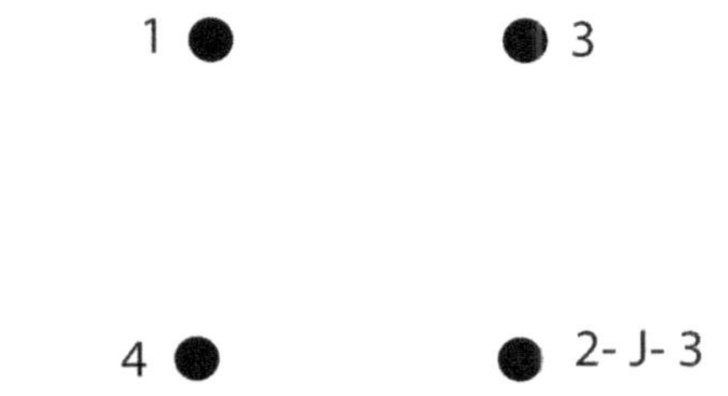

1 ●   ● 3

4 ●   ● 2-J-3

14

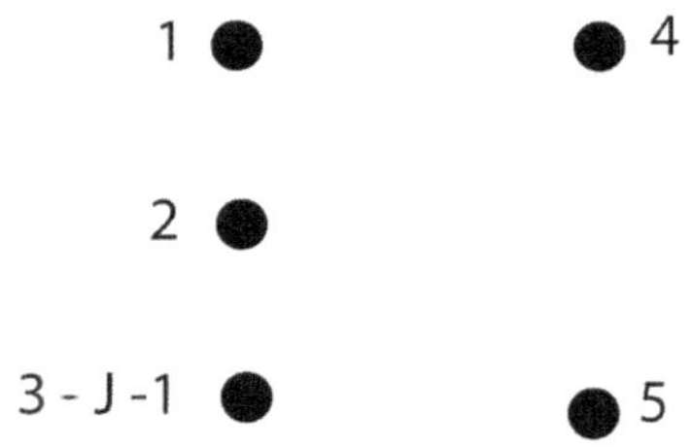

1 ●   ● 4

2 ●

3-J-1 ●   ● 5

15

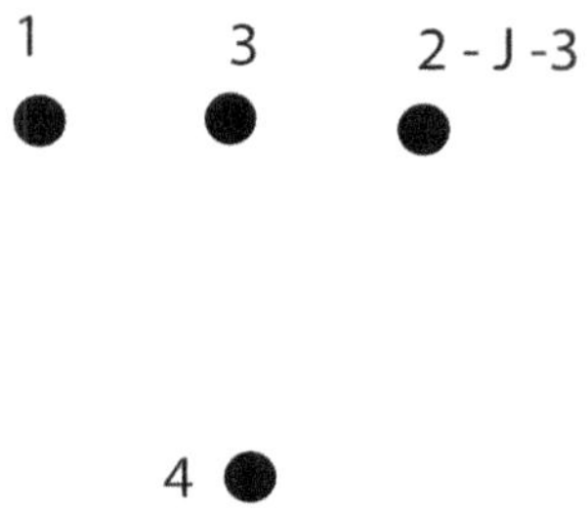

1 ●   3 ●   2-J-3 ●

4 ●

Pictorial Reasoning
© Gift Of Logic, Inc * Copying prohibited

Name _____   Date _____

CONNECT THE DOTS

Connect the dots shown below, starting from the highest number to the next lower number, and ending at the lowest. When you reach the lowest number, connect it to the highest number to complete the shape.

16

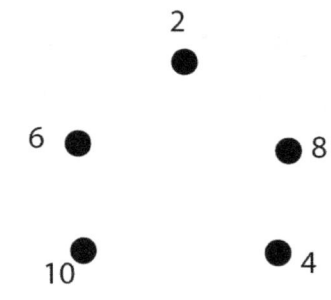

17

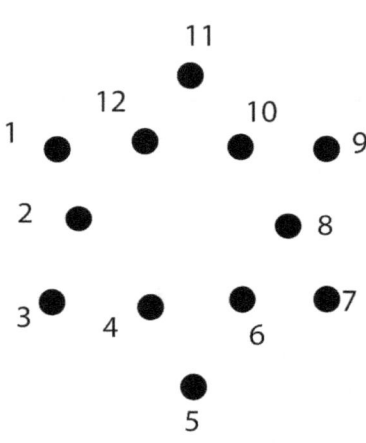

18

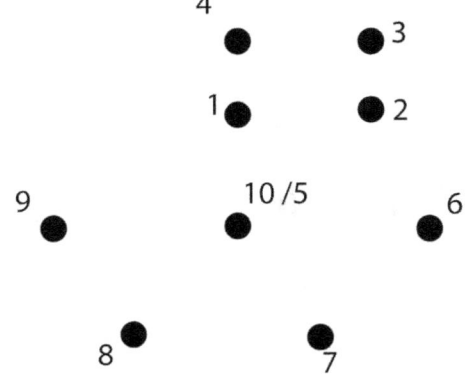

10/5 means that a dot can be assumed to be either 10 or 5.

Pictorial Reasoning   Answers-111   62
© Gift Of Logic, Inc * Copying prohibited

Name _____   Date _____

CONNECT THE DOTS - EVEN AND ODD

Connect the dots shown below, starting from the lowest number and ending at the highest number. Connect the odd numbers first, and then the even numbers.

19

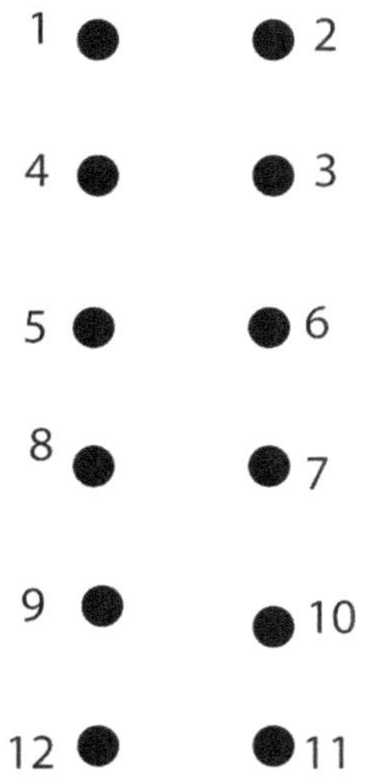

```
1 •      • 2
4 •      • 3
5 •      • 6
8 •      • 7
9 •      • 10
12 •     • 11
```

20

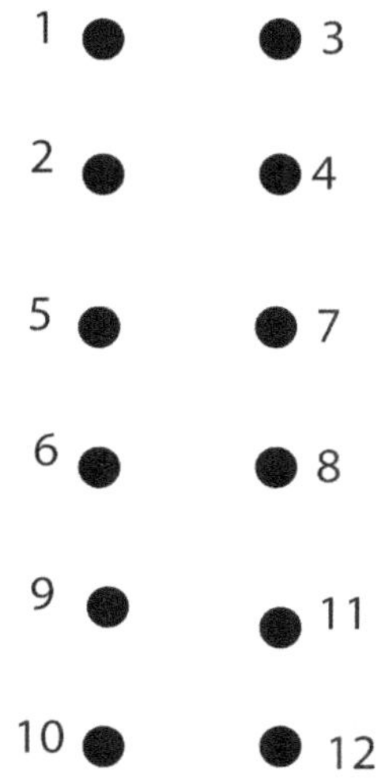

```
1 •      • 3
2 •      • 4
5 •      • 7
6 •      • 8
9 •      • 11
10 •     • 12
```

21

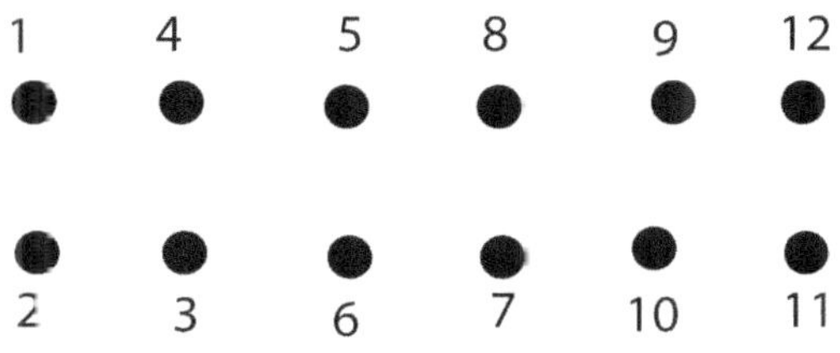

```
  1   4   5   8   9   12
  •   •   •   •   •   •

  •   •   •   •   •   •
  2   3   6   7   10  11
```

Pictorial Reasoning
© Gift Of Logic, Inc * Copying prohibited

Name _____   Date _____

MAZE

Solve the mazes shown below from Start to End.

1

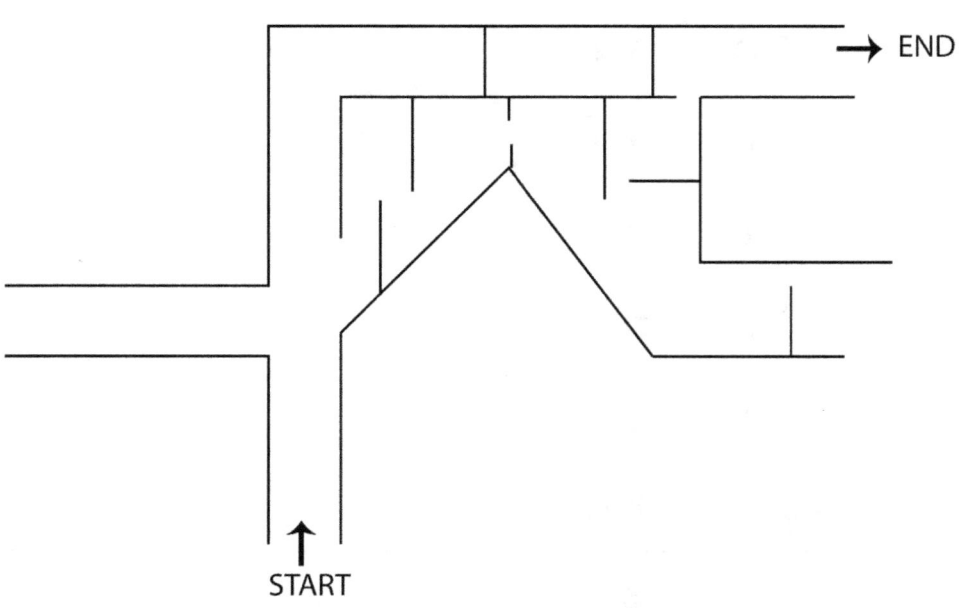

2

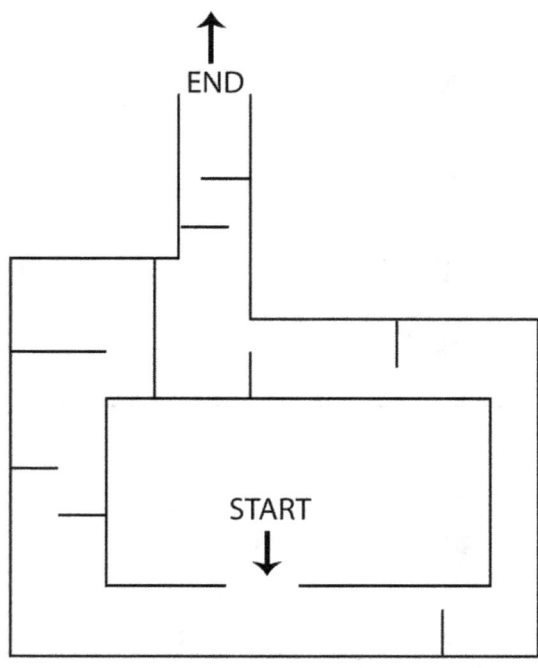

Name _____   Date _____

MAZE

Solve the mazes shown below from Start to End.

3

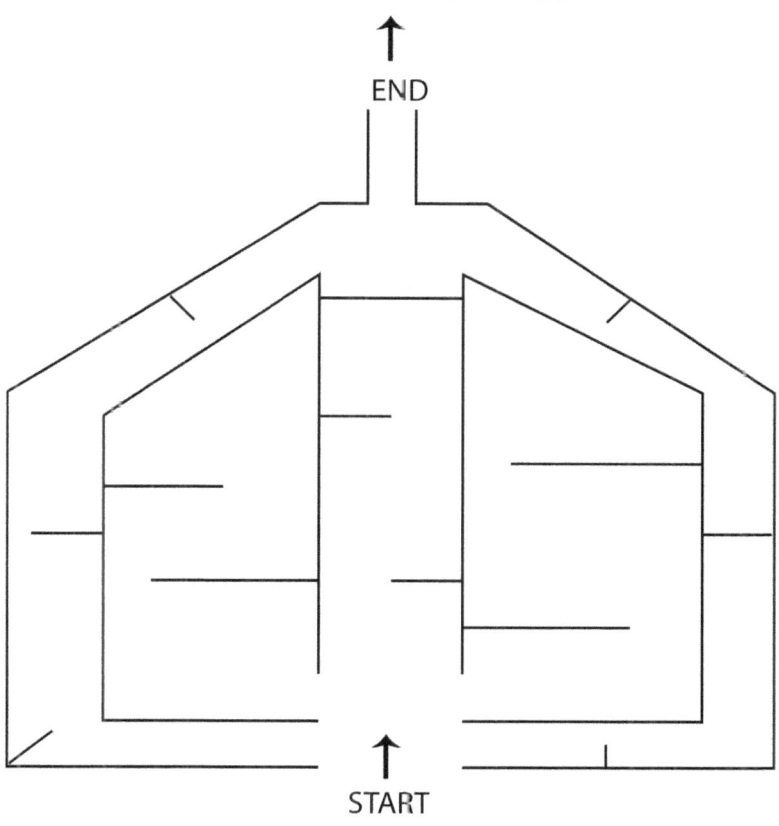

4

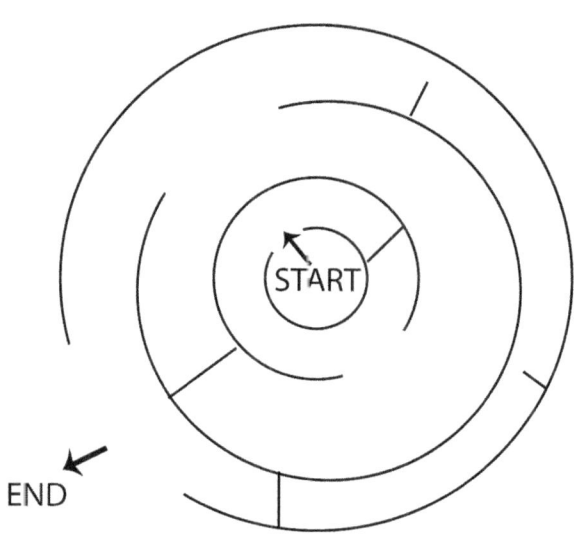

Name _____   Date _____

MAZE

Solve the mazes shown below from Start to End.

5

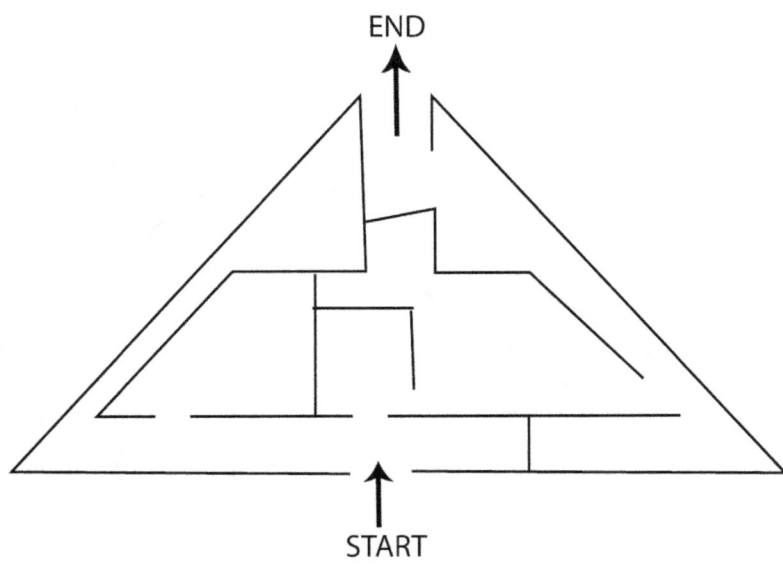

6

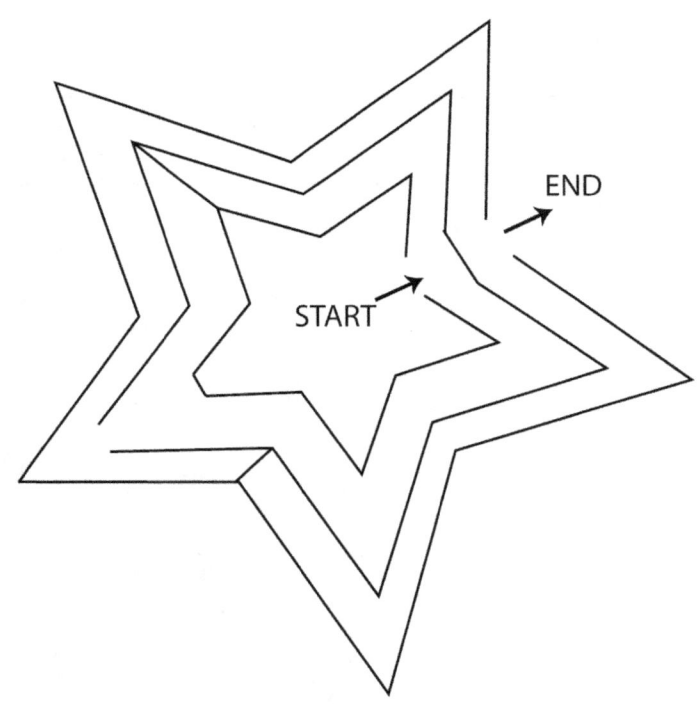

Name _____   Date_____

MAZE

Solve the mazes shown below from Start to End.

7

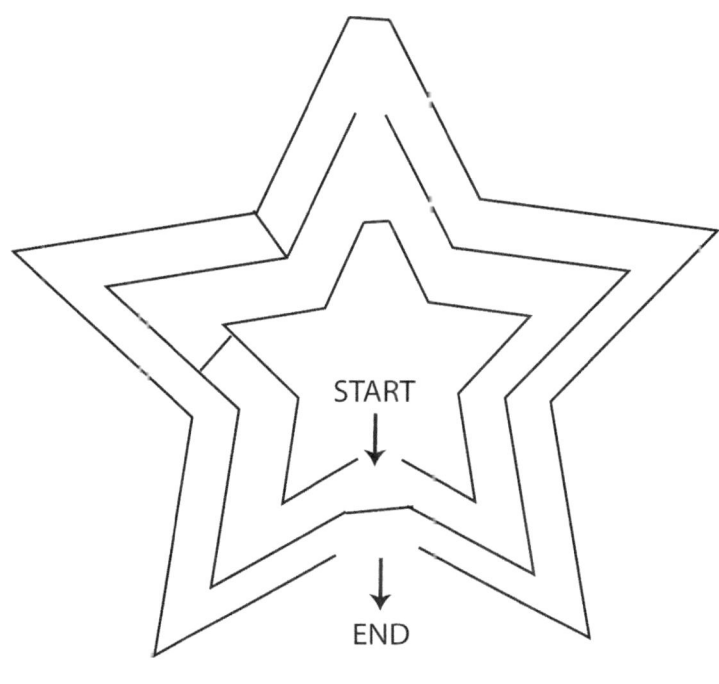

8

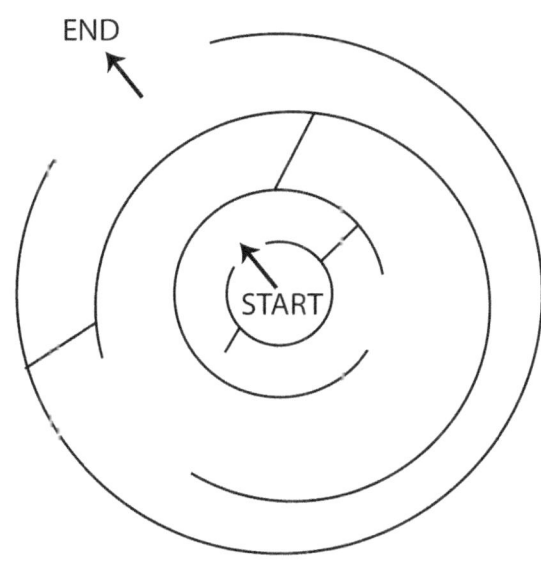

Name ——————————— Date ———————————

MAZE

Solve the mazes shown below from Start to End.

9

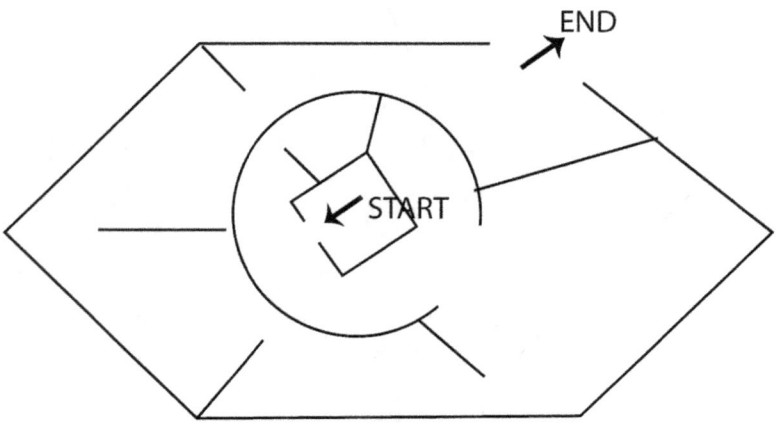

10

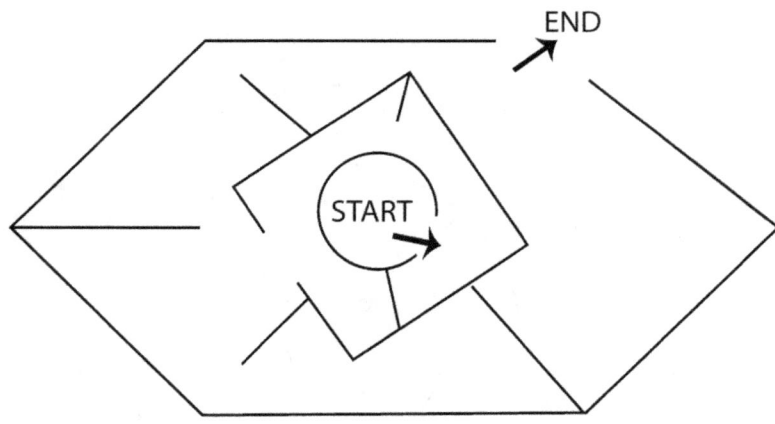

Name _____ Date _____

PICTURE SEQUENCE

Figure out the logic in the sequence of figures shown, and draw the next picture in the sequence that will continue the logic.

1

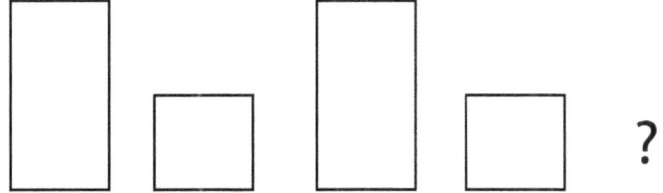

2

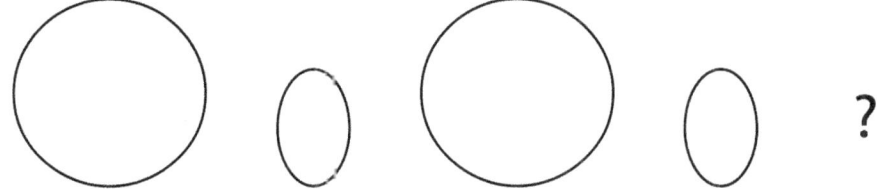

3

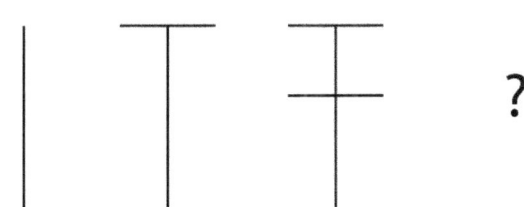

4

5

↓ → ↑ ?

Name _____ Date _____

PICTURE SEQUENCE

Figure out the logic in the sequence of figures shown, and draw the next picture in the sequence that will continue the logic.

6

7

8

9

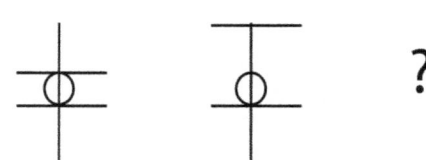

10

Pictorial Reasoning   Answers-113

© Gift Of Logic, Inc * Copying prohibited

Name _____   Date _____

PICTURE SEQUENCE

Figure out the logic in the sequence of figures shown, and draw the next picture in the sequence that will continue the logic.

11

12

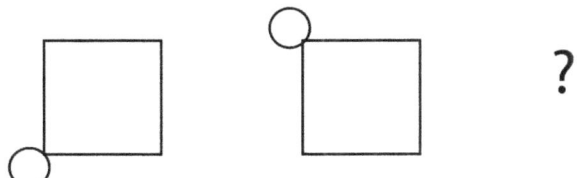

13

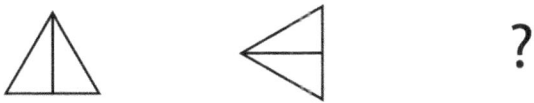

14

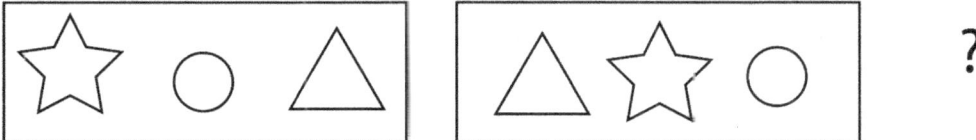

Pictorial Reasoning   Answers-113   71
© Gift Of Logic, Inc * Copying prohibited

Name _____   Date _____

ODD PICTURE

In each question below, find the odd picture and circle the answer.

1   A   B   C

2   A   B   C

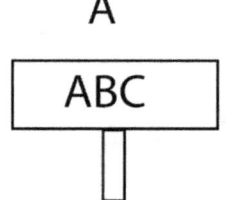 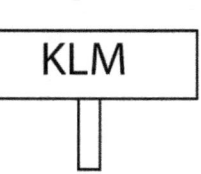 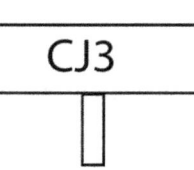

3   A   B   C

  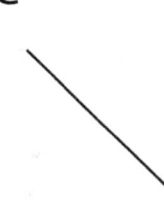

4   A   B   C

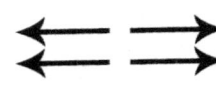  

Pictorial Reasoning   Answers-114
© Gift Of Logic, Inc * Copying prohibited

Name _____   Date _____

| ODD PICTURE |

In each question below, find the odd picture and circle the answer.

5 A   B   C

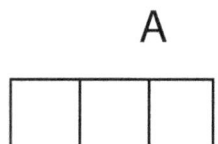 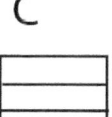

6 A   B   C

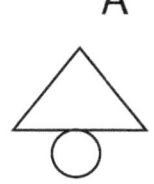  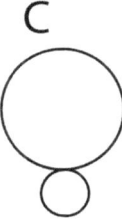

7 A   B   C

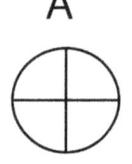  

8 A   B   C

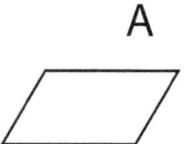 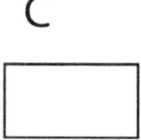

Pictorial Reasoning   Answers-114
© Gift Of Logic, Inc * Copying prohibited

Name ——————————— Date ———————————

ODD PICTURE

In each question below, find the odd picture and circle the answer.

9

A 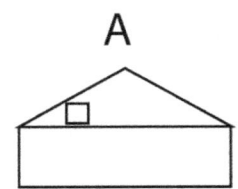   B 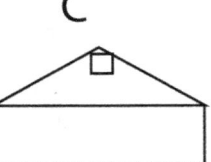   C

10

A 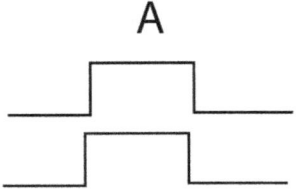   B   C

11

A    B 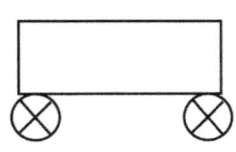   C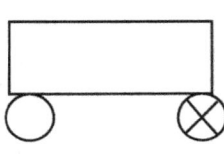

12

A    B 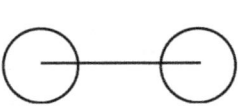   C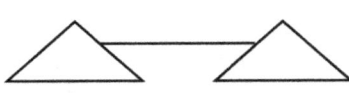

Pictorial Reasoning   Answers-114
© Gift Of Logic, Inc * Copying prohibited

SPOT THE DIFFERENCE

Spot the difference between the pictures shown and describe it.

1

2

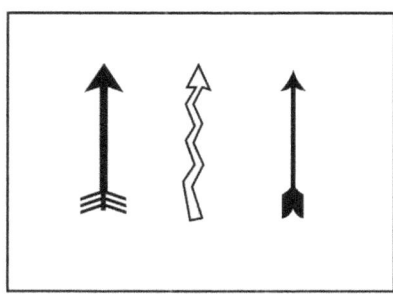 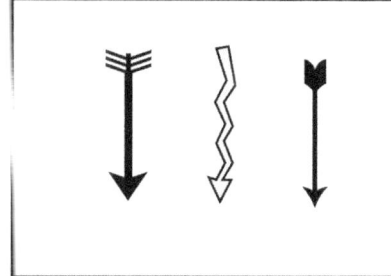

3

4

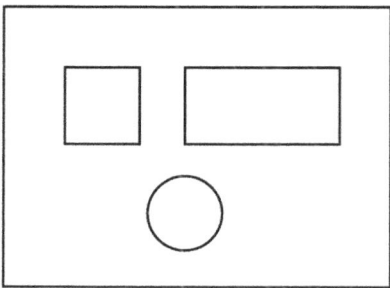 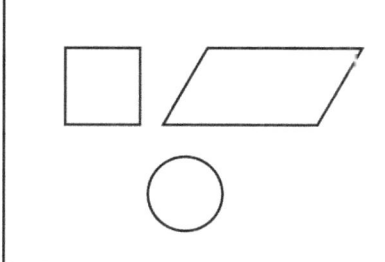

Name —————————————— Date ——————————————

SPOT THE DIFFERENCE

Spot the difference between the pictures shown and describe it.

5

6

7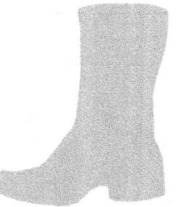

8

Name _____   Date_____

SPOT THE DIFFERENCE

Spot the difference between the pictures shown and describe it.

9

10

11

12

Pictorial Reasoning   Answers-115
© Gift Of Logic, Inc * Copying prohibited

ANSWERS

FINDING THE TRUTH

Q#	Answer	Reasoning
1	A-True	Every year has twelve months, January through December.
2	B-False	Not all months have thirty days. Only September, April, June, and November have thirty days.
3	B-False	Helicopters and Rockets can also travel in the air.
4	A-True	They lived on earth about 200 million years ago.
5	B-False	Cats, Parrots and several other animals and birds can be kept as pets.
6	B-False	We cannot see with our eyes closed.
7	B-False	Eiffel Tower is in the city of Paris in France.
8	A-True	Pluto is not considered as a planet in the Solar system anymore.
9	B-False	New York is closer to Atlantic Ocean, not the Pacific Ocean.
10	A-True	It is true that some car drivers are men.
11	B-False	Not all car drivers are men. Some car drivers are women.
12	A-True	Fire can be put out with water.

Answers
© Gift Of Logic, Inc * Copying prohibited

WORD ANALOGY

Q#	Answer	Reasoning
1	B	A bird moves using its wings. A fish moves using its fins.
2	B	Electricity flows through a wire. Water flows through a pipe.
3	A	Chair is used to sit. Treadmill is used to exercise.
4	B	It is hot inside an oven. It is cool inside a fridge.
5	A	You study in a class. You play in a park.
6	A	A room is closed with a door. A box is closed with a lid.
7	A	If we have a toothache, we must go to a dentist. Similarly, if we have a fever, we must go to a doctor.
8	A	Winning is success. Losing is defeat.
9	A	Increase means to raise. Decrease means to reduce.
10	B	We see with an eye. We hear with an ear.

Answers

SYNONYMS/ ANTONYMS

Q#	Answer	Reasoning
1	B-False	Noisy and quiet are antonyms.
2	A-True	Rude and arrogant mean the same.
3	A-True	To be slim is the same as being thin.
4	A-True	To be fat is the same as being heavy.
5	B-False	Friends do not quarrel.
6	B-False	Clean and tidy are synonyms.
7	B-False	To elevate something is to take it to a higher position, not bringing it down.
8	B-False	Scattering means dispersing, gathering means collecting.
9	B-False	When you help someone, you don't hinder them.
10	A-True	To permit someone is to allow them, not stop them.
11	A-True	Happy and sad are antonyms. To be happy is not the same as being sad.
12	B-False	Polishing and shining are synonyms. To say that polishing is not shining is a false statement.
13	B-False	Courteous and rude are antonyms.
14	A-True	Hate means dislike. To hate something is the same as not liking it.

Answers
© Gift Of Logic, Inc * Copying prohibited

AGREE-DISAGREE

1   Victor: Apples taste better ..   Victoria: Grapes taste better..
Answer: B) disagree with each other.

Reasoning: It is clear that Victor thinks apples taste better than grapes, but Victoria thinks otherwise.

---

2   Steve: Ants are smarter..   Emily: Elephants are smarter..
Answer: B) disagree with each other.

Reasoning: Steve says that ants are smarter than elephants. Emily says that elephants are smarter than ants. So, they both disagree with each other.

---

3   Daniel: When Josh got hurt..   Donna: The ambulance carrying..
Answer: B) disagree that the ambulance arrived quickly.

Reasoning: While Dan says that the ambulance arrived quickly, Donna disagrees by stating that "..the ambulance carrying the nurse could not find us quickly...". So, they both disagree that the ambulance arrived quickly to the scene.

---

4   Mom: Tony likes reading..   Dad: If Tony has to choose..
Answer: A) agree with each other.

Reasoning: Tony's mom says that Tony likes reading more than sports. His dad says that Tony will choose reading over sports. So, they both agree with each other.

Answers

© Gift Of Logic, Inc * Copying prohibited

INFERENCING

1   Tom was late.. Answer: B) Ryan went to the party before Tom.
<u>Reasoning:</u> It is clear from the statement that Tom was late, but Ryan was not. So, obviously Ryan would have gone to the party before Tom.

---

2   The diamond ring was in the jewel..
Answer: B) The diamond ring was found in the trash can.
   diamond ring → jewel case   *   jewel case → trash can
<u>Reasoning:</u> This can be easily inferred. Chaining the facts together will helps us to infer that the diamond ring was found in the trash can.

---

3   Amy's mom takes her ..
Answer: B) Amy's dad took her to the park today.
<u>Reasoning:</u> You can answer this question by breaking down the facts and writing them down.

   Mondays and Wednesdays - Mom
   Other days   - Dad

Now, the second statement clearly states that today is a Tuesday. Therefore, it can clearly be inferred that Amy's dad took her to the park today. Sorting the facts and writing them down will help you infer correctly.

---

4   Brian chased a thief..   Answer: C) Police Officer
<u>Reasoning:</u> Only police officers can arrest thieves, not others.

---

5   As Stacy stood in front..   Answer: A) an actress
<u>Reasoning:</u> Only actresses are likely to read drama scripts, not dentists.

Answers

INFERENCING

6

A bridge ..   Answer: C) an Engineer.

Reasoning: An Engineer is likely to give instructions to pour concrete for a bridge. Doctors and Lawyers are not likely to do this as their profession is different.

7

Troy dived.. Answer: B) a deep lake.

Reasoning: Since Troy dived several hundred feet, the place is likely a deep lake, not a shallow pond or a bathtub. A shallow pond and a bathtub do not run several hundred feet deep.

8
John was riding..   Answer: A) a remote place.

Reasoning: Since there was nobody around him, it is likely that he was in a remote place. If he was in the center of a city, it is very likely that someone would have been around him.

9
Neil jumped..

Answer: A) a tall building.

Reasoning: It is not likely that you have broken bones and head injuries if you jump from a small chair.

10

Martha watched TV.. Answer: B) between 6 PM and 8 PM.

Reasoning: Martha's TV was not stolen at 6 PM because she was watching it until 6 PM. It was not there at 8 PM, two hours later. So, we can infer that it was stolen between 6 PM and 8 PM.

Answers

INFERENCING

11 The Police rushed.. Answer: A) after the robber had left the train station.

<u>Reasoning:</u> The robber is in the train. The train left the station before the Police arrived. So, the Police arrived after the robber (who is in the train) had left the train station.

---

12 Jack and Jill were at the dental office ..
Answer: B) Jack and Jill got their teeth cleaned at the same time.
<u>Reasoning:</u> This is the only way both can come at the same time and leave in 30 minutes and still have their teeth cleaned.

---

13 Breaking down the facts and representing them in a tree-like diagram will help you reason correctly and quickly.

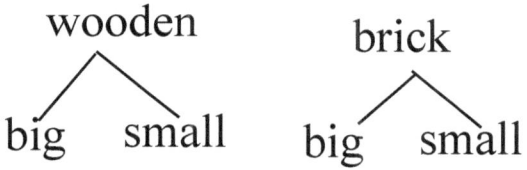

1) If a home is big, it must be a wooden home.
 Answer: B) False.
<u>Reasoning:</u> As you can see, a brick home can also be big.

2) If a home is made of bricks, it must be small.
Answer: B) False
<u>Reasoning:</u> As you can see, a brick home can also be big.

Answers

INFERENCING

14

Representing the facts in a tree-like diagram will help you to reason correctly and quickly.

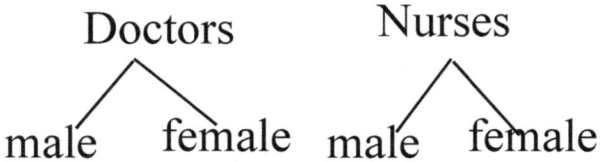

1) All doctors are female. Answer: B) False.
Reasoning: Some doctors are male as the figure clearly shows.

2) All nurses are male. Answer: B) False.
Reasoning: Some nurses are female as the figure clearly shows.

3) All females are doctors. Answer: B) False.
Reasoning: Some females are Nurses as well as the figure clearly shows.

Answers
© Gift Of Logic, Inc * Copying prohibited

DECISION MAKING

1 It would take ten minutes to..

Answer: A) Pine Street.

Reasoning: Facts clearly state that it would take longer to go to the Red river if you take the Elm Street. This means that taking the Pine street is the quickest way to go to the river.

---

2 The sooner you go to bed tonight..

Answer: B) 8 PM.

Reasoning: The given statement says that the sooner one goes to bed, the better they will do in the test. So, it is better to go to bed at 8 PM rather than at 10 PM.

DECISION MAKING

3 The red milk carton has less milk..

Answer: B) add more milk to the red milk carton.

Reasoning: Since the red milk carton has less milk than the blue milk carton, adding more milk to the red carton will bring the level of milk in it to the same level as that of the blue carton.

4 The baby elephant at the zoo..

Answer: B) lose some pounds

Reasoning: Since the zoo keeper wants it to weigh 150 pounds, the elephant must lose some pounds (weight) from its current weight of 200 pounds.

5 During the rainy season ...

Answer: B) keep more food in the shelves.

Reasoning: Since people buy extra food during the rainy season, the store keepers must keep more food in the shelves.

6 Kevin had saved 10 dollars..

Answer: A) More money must be added to the piggy bank.

Reasoning: Kevin spent 5 dollars on movies, which means he now has 10-5=5 dollars. So, more money (5 more dollars) must be added to the piggy bank to bring the balance to 10 dollars.

Answers
© Gift Of Logic, Inc * Copying prohibited

DECISION MAKING

7 Anita paid three dollars..

Answer: B) Return one dollar to Anita.

Reasoning: Since Anita paid one dollar more than the cost of the bread, the shop keeper must return one dollar back to Anita.

---

8 There are ten students in Mr. Gary's ..
Answer: B) Move one student from Mr. Gary's class to Mr. Roger's class.
Reasoning: Mr. Gary-10 students. Mr. Roger-8 students. So, if you move one student from Mr. Gary's class to Mr. Roger's class, Mr.Gary will lose one and Mr.Roger will gain one. So, both classes will have nine students.

---

9 Silvia has to attend .. Answer: A) before 9:30 AM.

Reasoning: Since the homework will take 30 minutes to do, if she starts before 9:30 AM, she can finish it before 10 AM and attend the party. If she starts the homework after 9:30 AM, she can finish her homework only after 10 AM and will not be able to attend the party.

---

10 The Science project must be submitted..
Answer: A) on or before July 29.

Reasoning: Since the project will take three days to complete, if she starts on July 29, she can finish it on July 31 and submit it on Aug 1. If she starts any day after July 29, then she will not be able to submit it on August 1.

DECISION MAKING

11 To be fully occupied..
Answer: B) Work on four 10 minute tasks and one 20 minute task.
Reasoning: One hour has 60 minutes. Four 10 minute tasks will consume 40 minutes. Add to this one 20 minute task and the total duration will be 60 minutes or 1 hour.

12 The short hand of a clock.. Answer: A) The long hand must be moved back. Reasoning: The long hand of the clock was ahead of itself at 3 PM and hence must be moved back to 12.

13 1) If she changed her pillow on Wednesday, when will she change it again? 1) Answer: B) Friday Reasoning: Since Gina likes to changer her pillow cover every other day, if she changed it on Wednesday, she will change it again on Friday.

2) If she changed her pillow cover on Wednesday this week, will she change it on Wednesday next week? Answer: B) No
Reasoning: The days when she will change pillows is shown below.
    this week: <u>Wednesday,</u> Friday
    next week Sunday, Tuesday, Thursday, Saturday
This week, Gina changed her pillow cover on Wednesday. So, she will change it again on Friday. After Friday, she will change it again on Sunday which will take her to the next week. After Sunday, she will change it on Tuesday, skip Wednesday and change it again on Thursday.

14 A school.. Answer: B) at 9 AM. Reasoning: The first fire drill must start at 9 AM. If it starts after 9 AM, three fire drills cannot be conducted.

DECISION MAKING

15

A special red bulb will not glow immediately..

Answer: A) 9:45 AM. <u>Reasoning:</u> If the bulb is turned on at 9:45 AM, then it will glow at 10:00 AM, fifteen minutes after it is turned on.

16

A horse is sent to the stage when the circus begins, and every five minutes thereafter. An elephant is also sent to the stage when the circus begins, but every ten minutes thereafter.

Answer: C) A horse only.

<u>Reasoning:</u> Although this appears to be a tough question, it can be easily solved if you write out the facts in a time line.

Write the time line of the circus and mark the animal that is sent to the stage. H represents a horse and E represents an elephant.

```
              <======= time ======>
circus-begins  5     10    15    20
     H         H     H     H     H
     E               E           E
```

From the time line, it is clear that after fifteen minutes, only a horse must be sent to the stage.

Answers
© Gift Of Logic, Inc * Copying prohibited

POSITIONING

1

| Red box | Green box | Blue box |

Three boxes shown above can hold balls of the colors that are shown in the box. Six balls of the colors shown below must be placed in the boxes.

Ball 1 - Green   Ball 2 - Red   Ball 3 - Blue
Ball 4 - Green   Ball 5 - Red   Ball 6 - Green

Answer the questions below.

1) How many balls can be placed in the red box?
Answer: A) 2
Reasoning: Ball 2 and Ball 5 are the two red colored balls that can be placed in the red box.

2) How many balls can be placed in the green box?
Answer: A) 3
Reasoning: Balls numbered 1, 4 and 6 are the three green balls that can be placed in the green box.

3) How many balls can be placed in the blue box?
Answer: A) 1
Reasoning: Only one ball, Ball 3 is blue in color.

Answers
© Gift Of Logic, Inc * Copying prohibited

POSITIONING

2

Jack and Jill must sit on the chairs shown above. Jack must sit in chair # 1.

Which one of the following seatings is correct?

A) Incorrect- Jack must sit in chair # 1, but he is sitting in chair # 2.

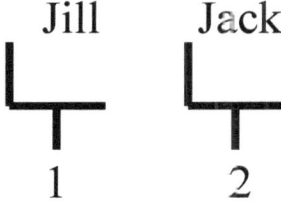

B) Correct - Jack is sitting in chair # 1. This is the correct seating arrangement.

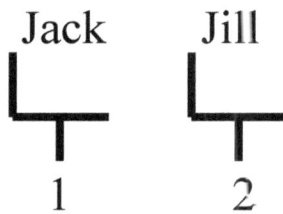

Answers

POSITIONING

3

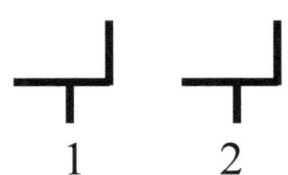

Mary and Nancy must be seated in the two chairs shown above. Mary can sit in either chair #1 or chair #2.

In the following seating arrangements, write the name of the person sitting in chair #1 and chair #2 next to the question mark.

In arrangement (A) below, Mary is seated in chair #1. In arrangement (B) below, Mary is seated in chair #2. She can sit in chair #1 or chair #2. The "or" is important to note.

A) ? Mary   ? Nancy

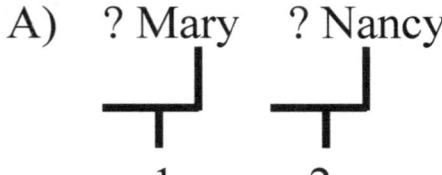

B) ? Nancy   ? Mary

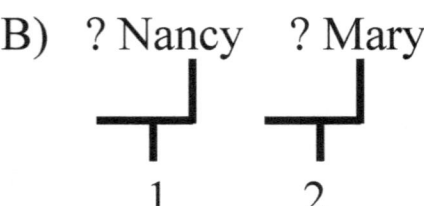

POSITIONING

4

Rick, Shelly, and Tina must be seated in the three chairs shown above. Shelly must sit next to Rick. This means that Shelly can sit to the left of Rick or to the right of Rick. Tina must sit in chair # 3.

How many seating arrangements are possible? Answer: Only two seating arrangements are possible. They are shown below.

A)   ? Rick   ? Shelly   Tina

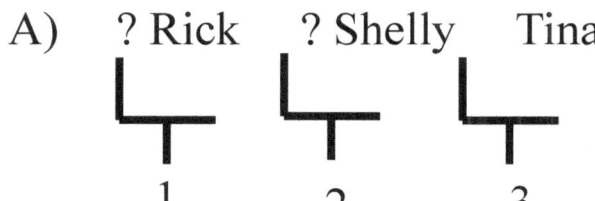

B) ? Shelly   ? Rick   Tina

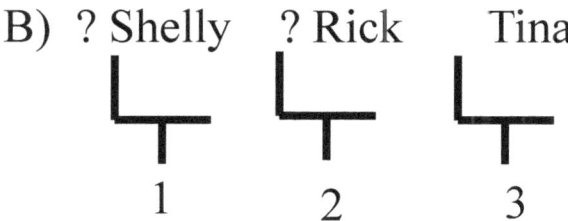

C) The valid seating arrangements are shown above.

   ?   ?   Tina

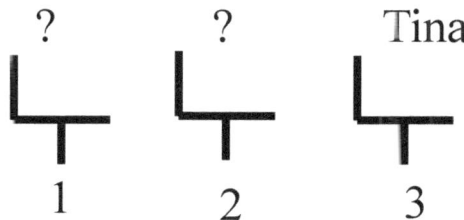

Answers

POSITIONING

5
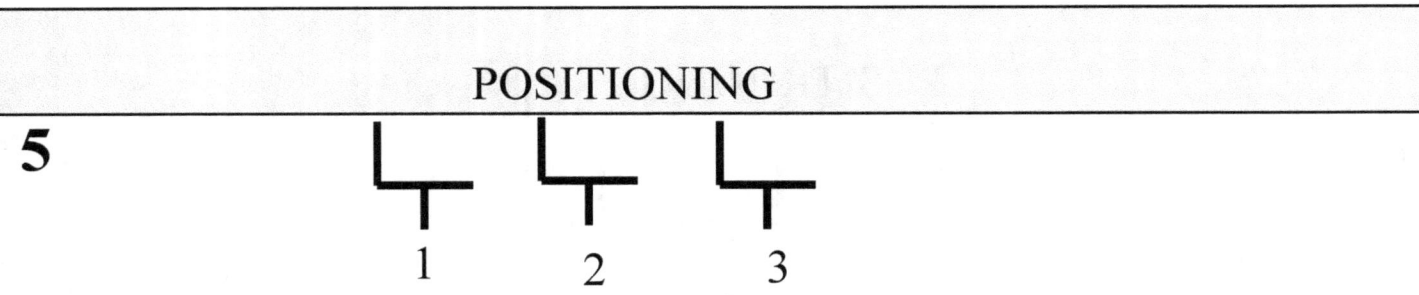

Emma, Frank, and Gary must be seated in the three chairs shown.
Frank can sit in the first chair or the last chair only.

If Frank sits on the first chair, where can Emma and Gary sit? Write their names in their possible positions below.

A)   Frank   ? Emma   ? Gary
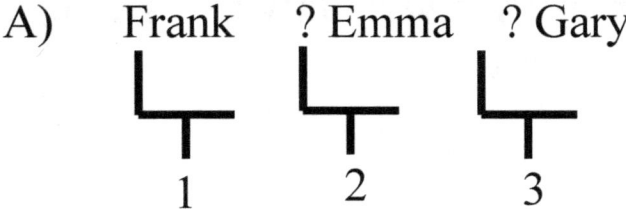

B)   Frank   ? Gary   ? Emma
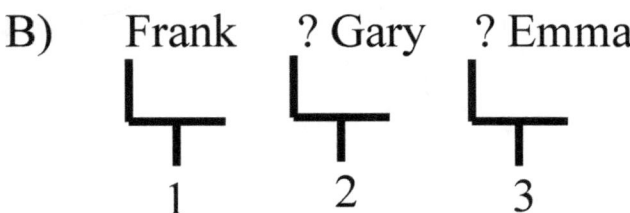

If Frank sits on the last chair, where can Emma and Gary sit? Write their names in their possible positions below.

A)   ? Emma   ? Gary   Frank
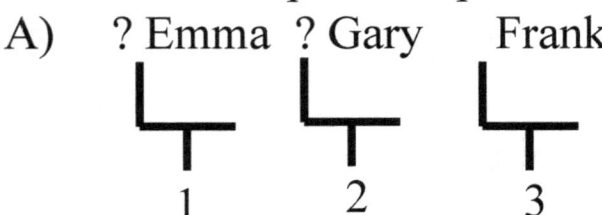

B)   ? Gary   ? Emma   Frank
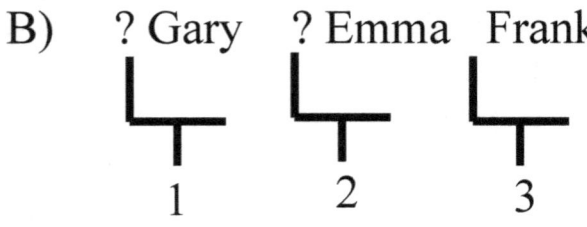

Answers

NUMERIC SUDOKU

1

	1	4	3
3		1	2
1	3		4
4	2	3	

2	1	4	3
3	4	1	2
1	3	2	4
4	2	3	1

2

1			2
2	3	1	4
4			3
3	2	4	1

1	4	3	2
2	3	1	4
4	1	2	3
3	2	4	1

3

2			3
1	3	2	4
3			1
4	1	3	2

2	4	1	3
1	3	2	4
3	2	4	1
4	1	3	2

Answers

© Gift Of Logic, Inc * Copying prohibited

NUMERIC SUDOKU

4

	3	1	
2	1	3	4
	4	2	
3	2	4	1

4	3	1	2
2	1	3	4
1	4	2	3
3	2	4	1

5

1	3	4	2
	2	1	
	4	3	
3	1	2	4

1	3	4	2
4	2	1	3
2	4	3	1
3	1	2	4

6

	3		1
4		3	
1	4	2	3
3	2	1	4

2	3	4	1
4	1	3	2
1	4	2	3
3	2	1	4

Answers

NUMERIC SUDOKU

7

	2	3	1
3	1		2
1		2	3
2		1	

4	2	3	1
3	1	4	2
1	4	2	3
2	3	1	4

8

3	1	4	2
4			3
2	4		1
1	3	2	

3	1	4	2
4	2	1	3
2	4	3	1
1	3	2	4

9

	4	1	3
1		2	4
3		4	1
4	1		2

2	4	1	3
1	3	2	4
3	2	4	1
4	1	3	2

Answers

NUMERIC SUDOKU

10

4	3	1	2
2		3	4
1	4		3
3	2	4	

4	3	1	2
2	1	3	4
1	4	2	3
3	2	4	1

11

4	1	3	
3	2		4
2		4	1
	4	2	3

4	1	3	2
3	2	1	4
2	3	4	1
1	4	2	3

12

1	2	4	
3	4		2
4		2	1
2		3	4

1	2	4	3
3	4	1	2
4	3	2	1
2	1	3	4

Answers
© Gift Of Logic, Inc * Copying prohibited

NUMERIC SUDOKU

13

4	1	2	3
2	3		1
3		1	2
	2	3	4

4	1	2	3
2	3	4	1
3	4	1	2
1	2	3	4

14

2		1	3
3		2	4
1	3		2
4	2		1

2	4	1	3
3	1	2	4
1	3	4	2
4	2	3	1

15

4	1		3
3	2		4
2	3	4	
1	4	3	

4	1	2	3
3	2	1	4
2	3	4	1
1	4	3	2

Answers
© Gift Of Logic, Inc * Copying prohibited

NUMERIC SUDOKU

16

1	3		4
	4	3	1
4	2		3
3	1	4	

1	3	2	4
2	4	3	1
4	2	1	3
3	1	4	2

17

1	2	3	4
	4	1	2
4		2	1
2	1		3

1	2	3	4
3	4	1	2
4	3	2	1
2	1	4	3

18

4	1	3	
2			1
1	4		3
	2	1	4

4	1	3	2
2	3	4	1
1	4	2	3
3	2	1	4

Answers 102
© Gift Of Logic, Inc * Copying prohibited

NUMERIC SUDOKU

19

2	1	4	3
4		2	1
3	2		4
1		3	2

2	1	4	3
4	3	2	1
3	2	1	4
1	4	3	2

20

	3	2	1
1	2		3
2	1		4
3	4	1	

4	3	2	1
1	2	4	3
2	1	3	4
3	4	1	2

Answers

ALPHABETIC SUDOKU

1

	A	D	C
C		A	B
A	C		D
D	B	C	

B	A	D	C
C	D	A	B
A	C	B	D
D	B	C	A

2

A	D	C	
B	C		D
D		B	C
	B	D	A

A	D	C	B
B	C	A	D
D	A	B	C
C	B	D	A

3

	D		C
A	C	B	D
	B		A
D	A	C	B

B	D	A	C
A	C	B	D
C	B	D	A
D	A	C	B

Answers

ALPHABETIC SUDOKU

4

	C	A	B
B	A		D
	D		C
C	B	D	A

D	C	A	B
B	A	C	D
A	D	B	C
C	B	D	A

5

	C	D	
D	B	A	C
B	D	C	A
	A	B	

A	C	D	B
D	B	A	C
B	D	C	A
C	A	B	D

6

B		D	
D	A	C	B
A		B	
C	B	A	D

B	C	D	A
D	A	C	B
A	D	B	C
C	B	A	D

Answers

ALPHABETIC SUDOKU

7

	B		A
C	A	D	B
A	D	B	C
	C		D

D	B	C	A
C	A	D	B
A	D	B	C
B	C	A	D

8

A	D	B	C
B		A	D
D	B		A
C		D	B

A	D	B	C
B	C	A	D
D		C	A
C	A	D	B

9

		A	C
A	C		
C	B		A
D	A	C	B

B	D	A	C
A	C	B	D
C	B	D	A
D	A	C	B

Answers

ALPHABETIC SUDOKU

10

D	C	A	B
	A	C	D
A			C
C	B		

D	C	A	B
B	A	C	D
A	D	B	C
C	B	D	A

11

	A	C	B
C		A	D
B			A
A	D	B	

D	A	C	B
C	B	A	D
B	C	D	A
A	D	B	C

12

	B	D	C
C	D	A	
D			A
B	A	C	D

A	B	D	C
C	D	A	B
D	C	B	A
B	A	C	D

Answers   © Gift Of Logic, Inc * Copying prohibited

ALPHABETIC SUDOKU

13

C	D	B	A
B		D	
D	C	A	B
A		C	

C	D	B	A
B	A	D	C
D	C	A	B
A	B	C	D

14

	D	A	C
C	A		D
	C	D	B
D	B		A

B	D	A	C
C	A	B	D
A	C	D	B
D	B	C	A

15

	B	A	
C	A	B	D
B		D	A
A	D		B

D	B	A	C
C	A	B	D
B	C	D	A
A	D	C	B

Answers

ALPHABETIC SUDOKU

16

A	C	B	
	D		A
D	B	A	C
	A	D	

A	C	B	D
B	D	C	A
D	B	A	C
C	A	D	B

17

	B		D
C	D		B
D			A
B	A	D	C

A	B	C	D
C	D	A	B
D	C	B	A
B	A	D	C

18

	A	C	B
B		D	A
A	D		C
C	B	A	

D	A	C	B
B	C	D	A
A	D	B	C
C	B	A	D

Answers
© Gift Of Logic, Inc * Copying prohibited

ALPHABETIC SUDOKU

19

B		D	
D	C		A
C		A	D
A	D		B

B	A	D	C
D	C	B	A
C	B	A	D
A	D	C	B

20

	C	B	A
A		D	
B	A		D
C		A	B

D	C	B	A
A	B	D	C
B	A	C	D
C	D	A	B

Answers

CONNECT THE DOTS

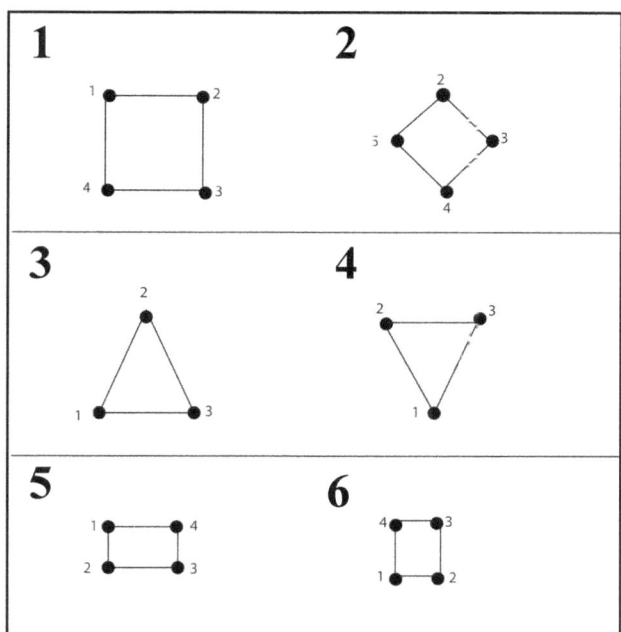
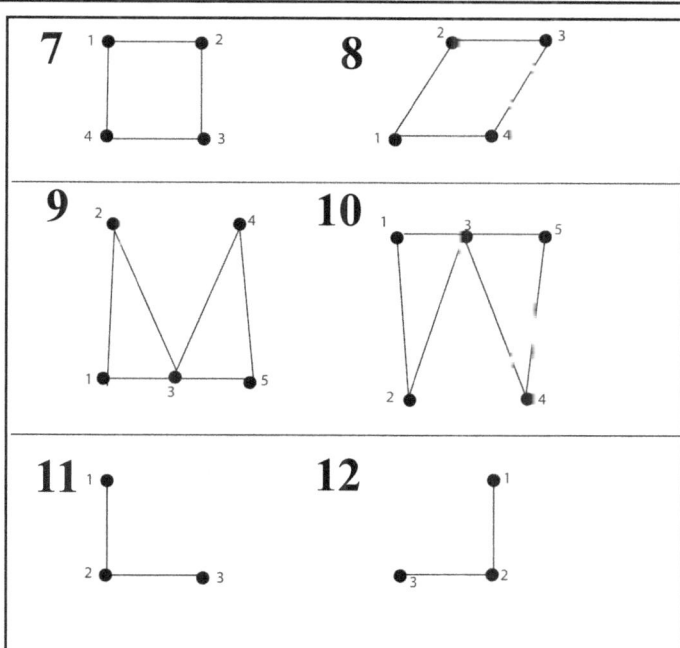
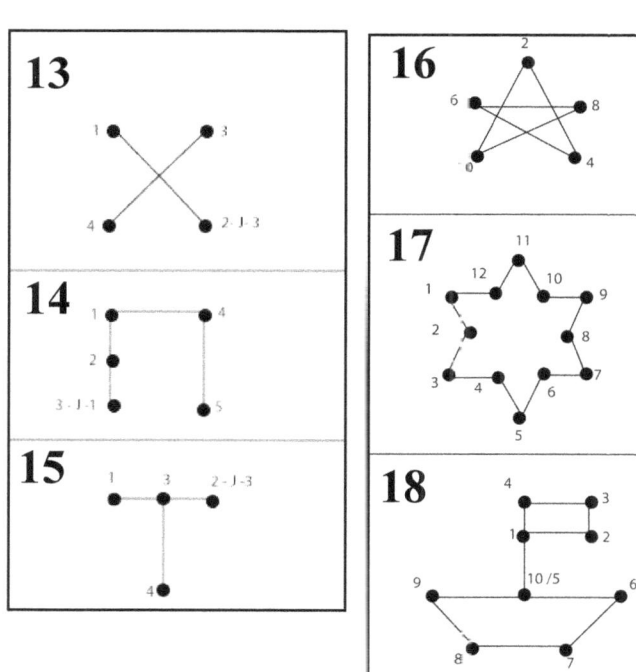
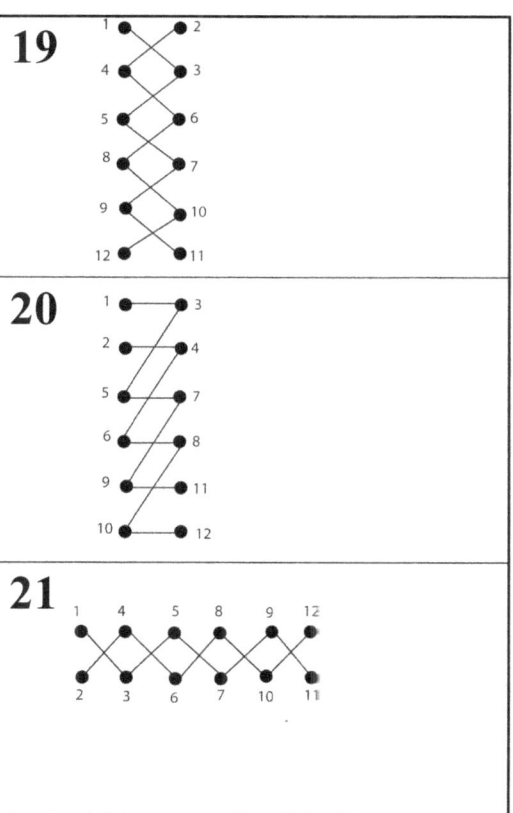

Answers 111
© Gift Of Logic, Inc * Copying prohibited

MAZES

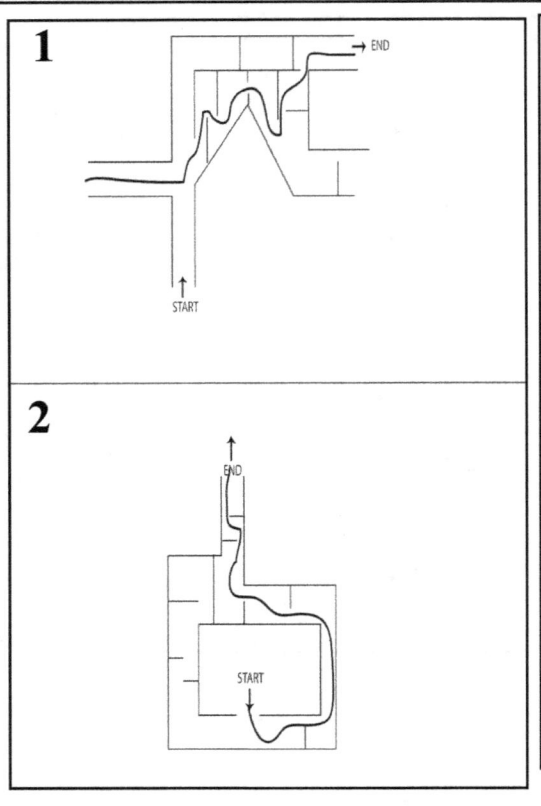
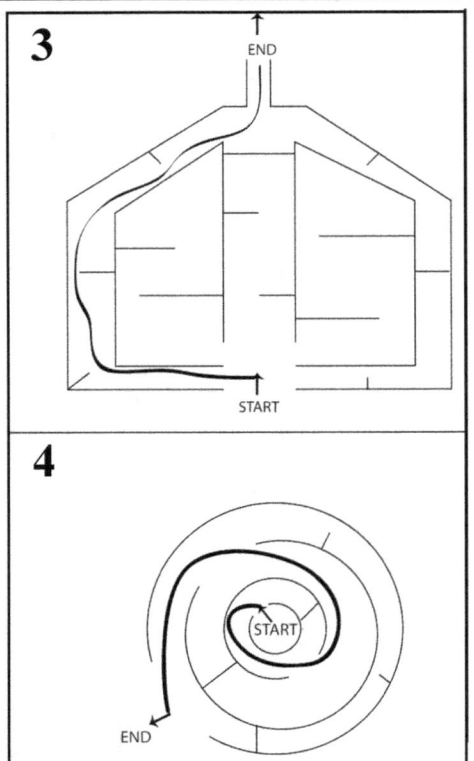
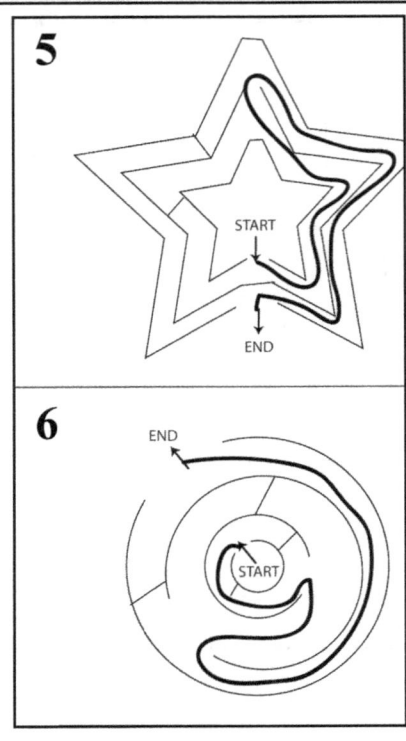
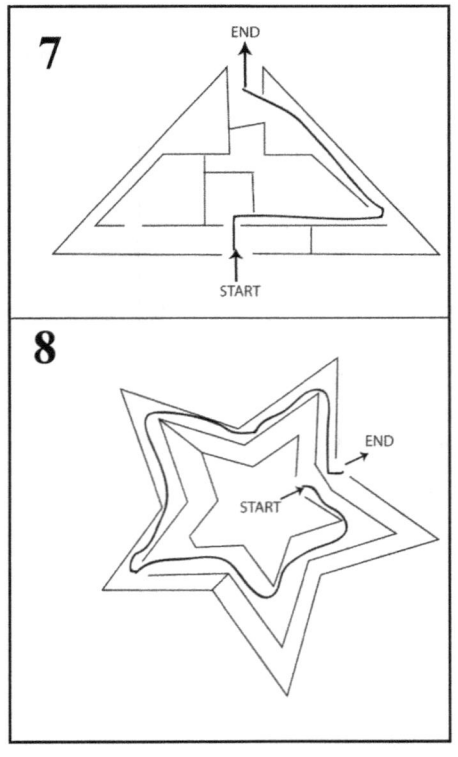
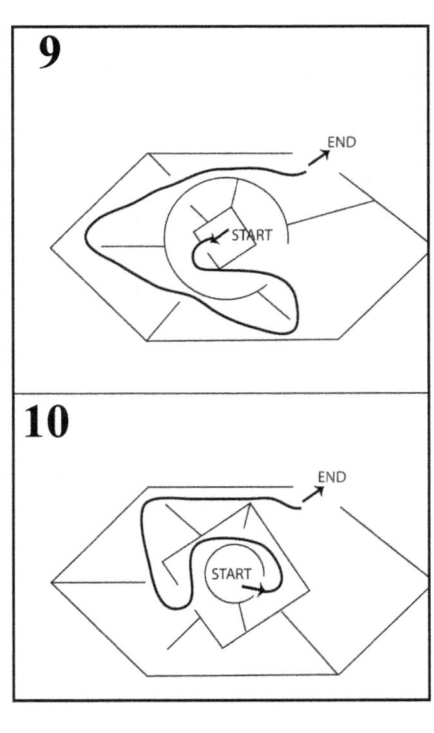

Answers

112

PICTURE SEQUENCE

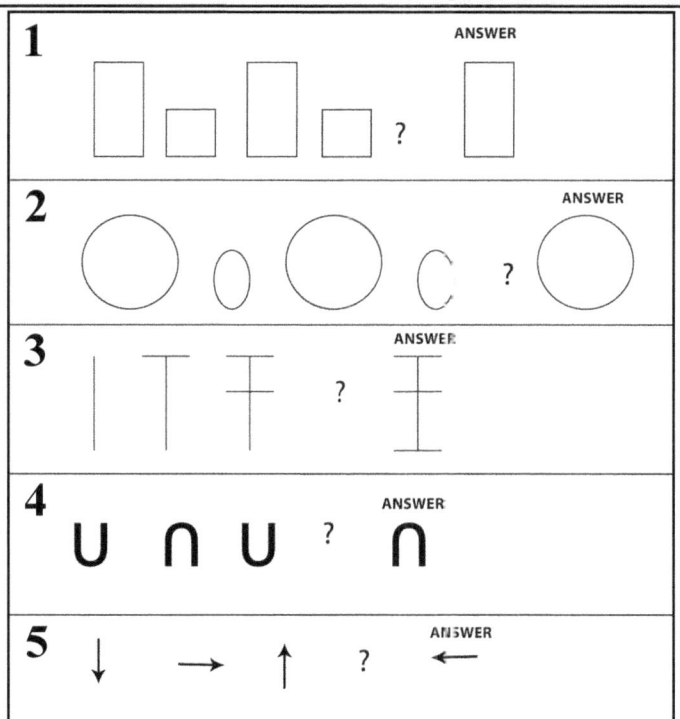
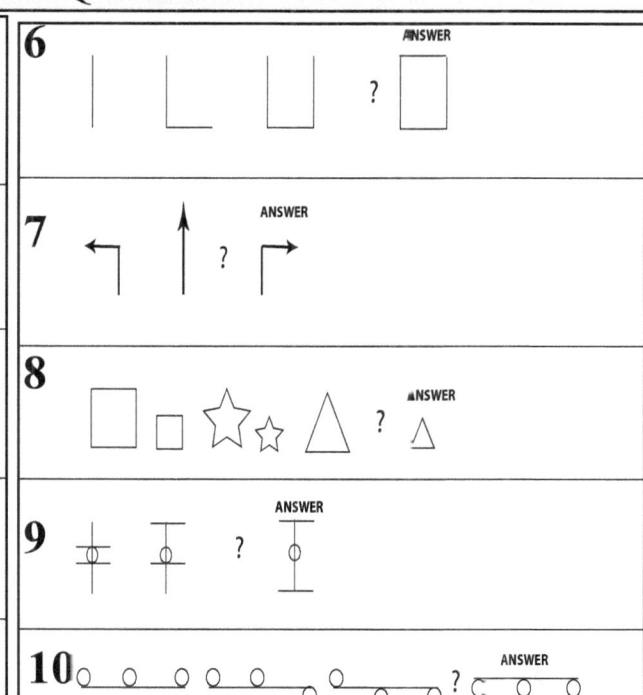
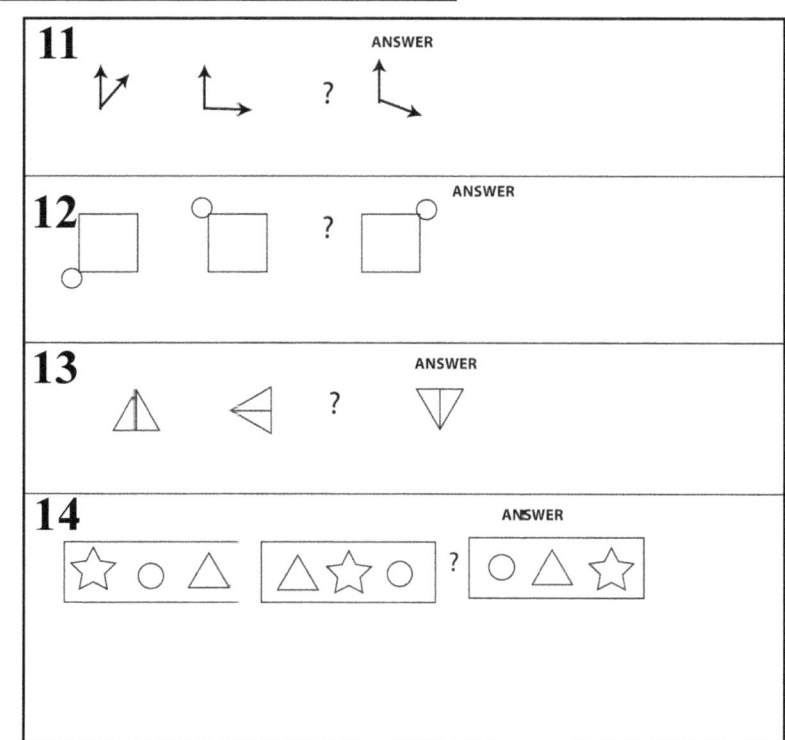

Answers 113
© Gift Of Logic, Inc * Copying prohibited

ODD PICTURE

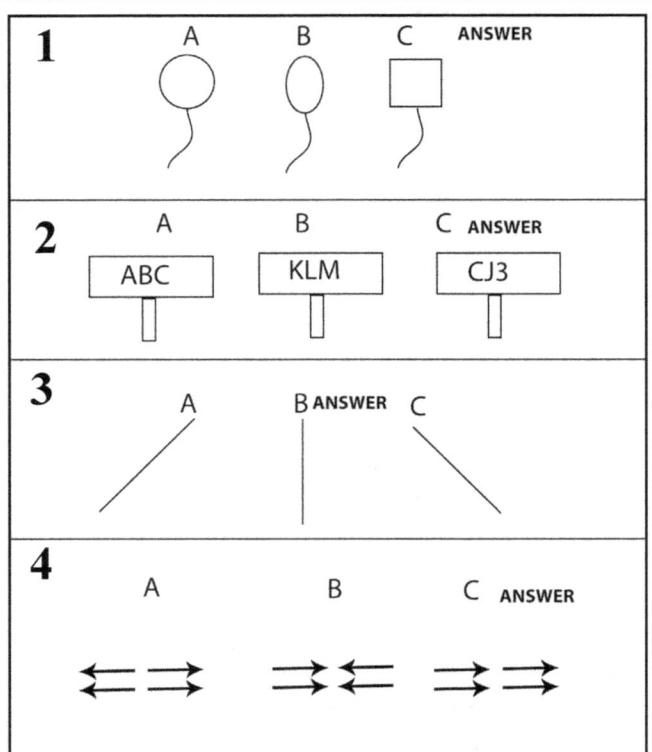
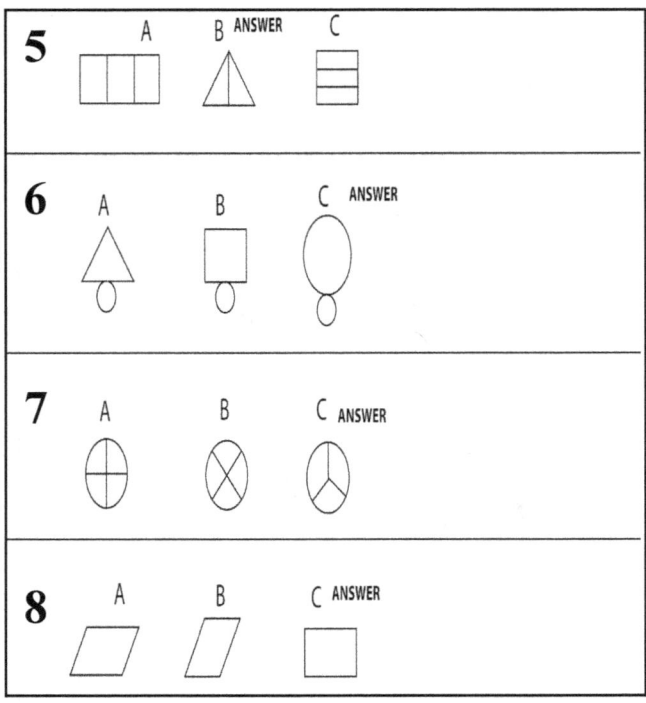
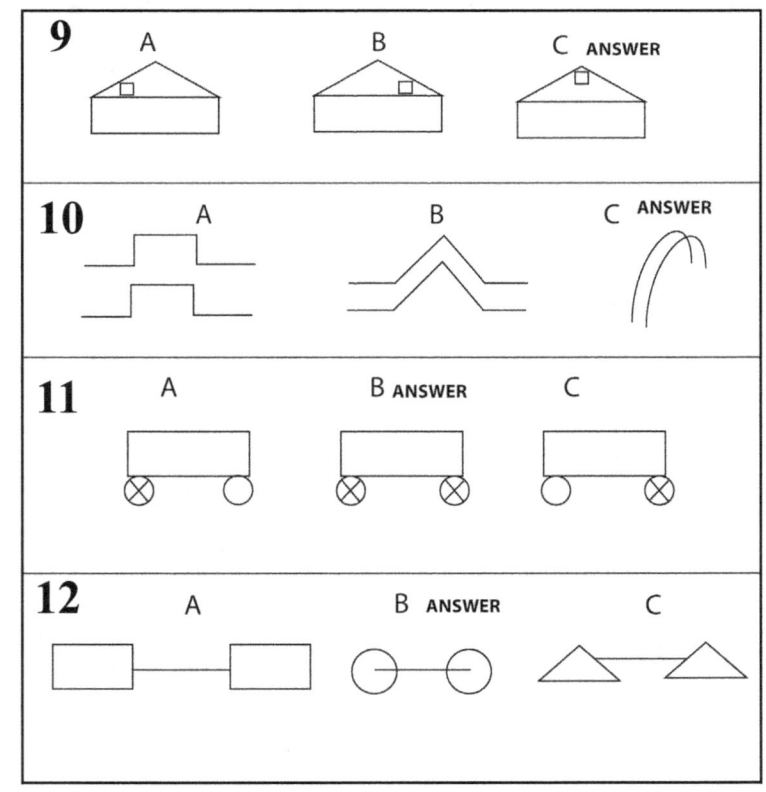

SPOT THE DIFFERENCE

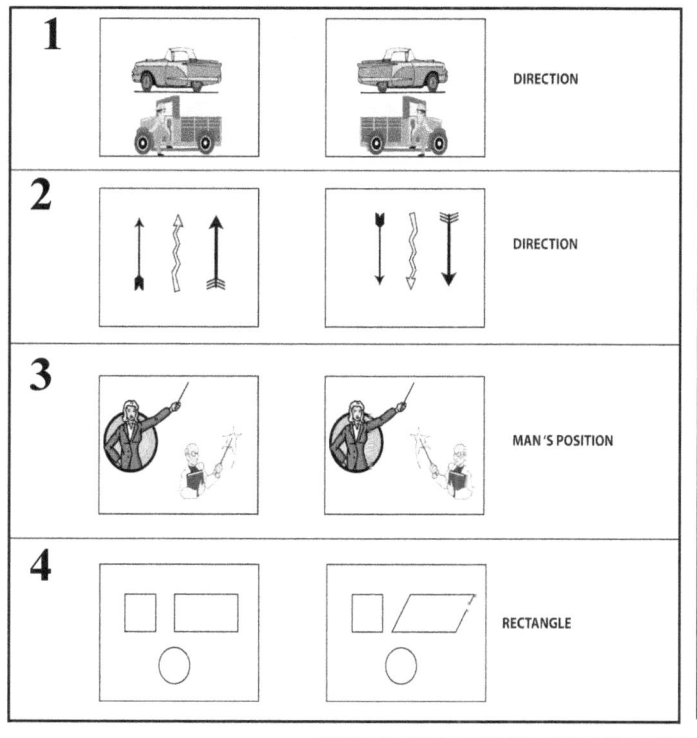

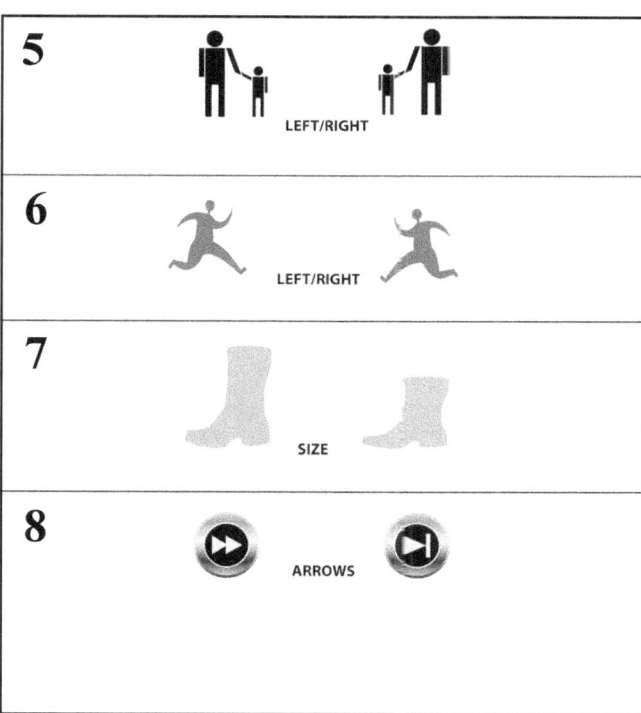

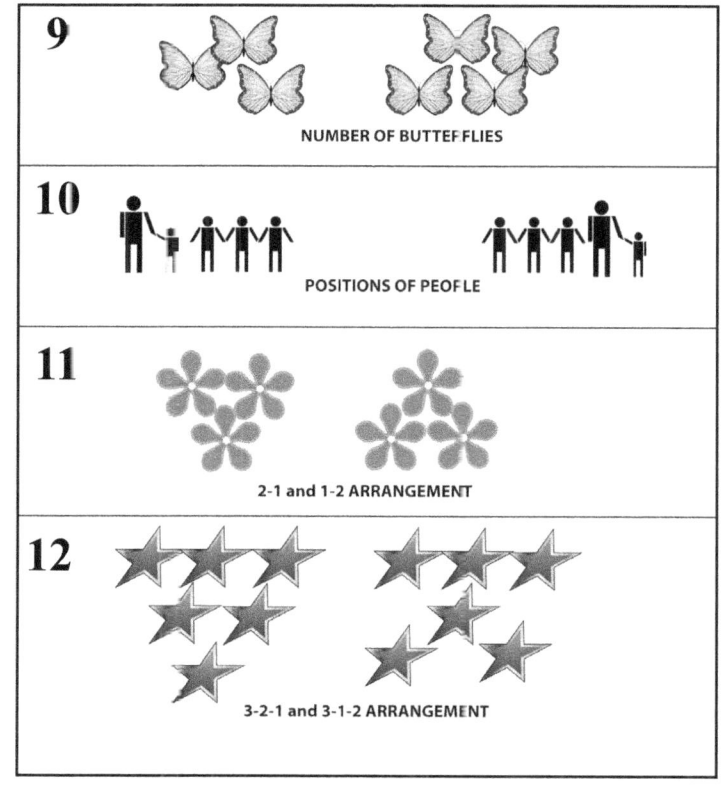

Answers

NOTES

NOTES

NOTES

NOTES